THE MAN WHO MADE CHINA
A LITERATE NATION –

ZHOU YOUGUANG

FATHER OF THE PINYIN WRITING SYSTEM

By Mark O'Neill

THANKS AND ACKNOWLEDGEMENTS

When I lived in Beijing, my home and office were not far from the apartment of Zhou Youguang. After he retired at the end of 1991, he remained famous among the scholars and intellectuals we knew. His nickname was 周百科, Zhou Baike – Encyclopaedia Zhou; this was because of his extraordinary knowledge of many subjects. We saw evidence of this in the many interviews he gave to newspapers and magazines, Chinese and foreign, on a wide range of topics. He published a new book every three years. So I had a good idea of his opinions on many subjects, which he expressed clearly and eloquently. To my shame, I did not go to visit him, as I should have. He was warm and friendly and welcomed visitors of all nationalities to his small apartment. My only excuse is that, like a mouse in a cage, we were busy pursuing the news of the day – but a two-hour meeting with Professor Zhou would have told you more about China, past and present, than most of the articles we wrote. What made him so unusual was that he was willing to speak the truth. During his life, both Nationalist and Communist parties invited him to join; but, like Hu Shih, the intellectual he most admired, he never joined a political party. Both men wanted to keep their status as an independent public intellectual who could speak their mind on the great issues of the day. This is what makes his biography so worth writing. He lived for 111 years, through four dynasties, two wars and the Cultural Revolution; very few others were witnesses to history as he was. His honesty, objectivity, humour and encyclopaedic memory make this book worth reading.

His prolific output means that he has left a rich legacy for his biographers. We have his books, articles and the many interviews he gave. For this biography, I used three books in particular. One was an oral autobiography published by the Chinese University of Hong Kong in 2015; over 525 pages, Zhou covers all the periods of his life in meticulous and humorous detail. The second was a book written by a Beijing scholar, Chen Guangzhong, and published in Taiwan in 2012; he conducted many interviews with Dr Zhou. The third was interviews given by Zhou to journalists and scholars between 2005 and 2014; it was published in Hong Kong in 2015. You will find the details of these three books in the sources at the end of the chapters. They were my primary raw material for this biography. We express our sincere thanks to the authors, scholars and journalists who wrote and compiled these books – and, above all, to Dr Zhou himself for talking so frankly and in such detail. The wealth and clarity of his memory, even after he was 100 years old, were remarkable. We must also sincerely thank Madame Mao Xiaoyuan, his niece; she saw a great deal of him during his retirement years. She sent us excellent photographs and articles she wrote about him; these provided valuable material.

We thank others for their contribution, including Victor Mair, Professor of Chinese Language and Literature at the University of Pennsylvania, and Kang-I Sun Chang, Professor of East Asian Languages and Literatures at Yale University. Professor Chang provided excellent photographs, as did

Melinda Ye Minlei, Editorial Manager of the Chinese University of Hong Kong Press. We also thank Ambassador Paul Chang Kuobao of Taiwan for his revisions to the section on Taiwan in Chapter Eleven. We thank Helen Wu Weiran, our good friend in Shanghai. She provided excellent photographs and liaised with people in the mainland. Covid restrictions made it impossible for me to go there.

We should add a few words about the Romanisation of Chinese names and places used in this book. In 1958, the PRC government adopted Hanyu Pinyin as the national standard for Romanised spelling. Since then, as the book explains, it has become the international standard. But not everyone uses it. The wife of Zhou Youguang was Zhang Yunhe, but two of her sisters were Chang Yuan-he and Chang Ch'ung-he. They moved to the United States; these were their legal names in English. Many residents of Taiwan and Hong Kong and overseas Chinese use non-Pinyin spellings. In 2009, the Taiwan government adopted Pinyin as its standard – but has retained the previous spellings of major cities, such as Taipei, Kaohsiung and Keelung, because they are well established around the world. Our rule in this book is to use the spelling chosen by the person or institution we are writing about.

In the early 1950s, Zhang Yunhe and Zhou Youguang were in Suzhou Park.

The 1960s, Zhou Xiaoping, Zhou Heqing, Zhou Youguang, Zhang Yunhe, He Shixiu. (From left to right).

July 25, 1999, Zhang Yunhe's birthday.

Zhou Youguang was in his study.

CONTENTS

INTRODUCTION

The Man Who Made China A Literate Nation

When young Mainland children arrive in primary school, they go to a Chinese class. The professor teaches them not characters but Pinyin, an alphabet using Roman letters with four different tones. Once they have mastered it, then the professor starts to teach them characters; thanks to the Pinyin next to each character, they can read it. Without it, they would not know how to pronounce them.

It is this system – full name Hanyu Pinyin, meaning Chinese phonetics – which has turned China into a literate nation. It was introduced in 1958, when 80 per cent of the population was illiterate. Since then, a billion people have learnt Pinyin and the rate of illiteracy has fallen to below 10 per cent.

The invention of Pinyin has also been a godsend for the millions of foreigners who learn Chinese. In numbers of people it has made literate, it is the greatest achievement in linguistics in human history.

The man most responsible for Pinyin is Zhou Youguang, director of the department in the Chinese Character Reform Commission (CCRC) instructed in 1955 to create a new romanisation system.

Several phonetic systems already existed; some were created by foreign missionaries who needed to master Chinese to spread the gospel to the world's most populous nation. But Zhou – and the leaders of the government – believed that none of the existing systems was easy nor simple enough. It took him and his colleagues three years of intense work. In the middle, the Soviet Union, China's "Big Brother" at the time, pressed them to use Cyrillic letters rather than Roman ones. No, said Zhou, Roman letters were the global standard and understood by millions of overseas Chinese around the world.

The task was extremely difficult. Chinese has more than 100,000 characters. Secondary school students are expected to master 3,500, and university graduates 5,000-6,000. The greatest obstacle is the fact that hundreds of characters have the same sound; how to romanise them, when the Latin alphabet has only 26 letters?

In history, Chinese was the written language of several Asian countries, including Korea, Japan and Vietnam. All three took a different route than China did. Each decided that the characters were simply too difficult for their people to learn and they had to find an alternative.

In the 15th century, King Sejong the Great of Korea commissioned a group of scholars to create an alphabet suited to the spoken language used by his people; they invented Hangul with 14 consonant letters and 10 vowel letters. Today it is the official written language of both South and

North Korea. Since it was created by scholars, it was easy to learn and entirely logical.

In Vietnam, Portuguese, Italian and French missionaries in the 17th century created a Latin script to replace the characters. It is the basis of the Vietnamese written language used today. Japan uses three alphabets: Chinese characters; one called hiragana developed from the 7th century AD; and another called katakana, mainly for imported words. The latter two are phonetic, not pictograms. Once you have learnt them, you can pronounce the words.

After the Meiji Restoration in 1868, Japan had an intense debate over its language. Was it an obstacle to modernisation and learning the new terms in science, industry and armaments needed to catch up with the West? One scholar even proposed that they abandon Japanese completely and adopt French instead. In the end, the government decided to retain the three alphabets; they are still in use today.

China did not follow the example of Korea, Vietnam and Japan. It has maintained characters as its written script. It is the most difficult language in the world to learn, because each character is distinct and must be learnt individually. That is why the rate of literacy was so low – 10-15 per cent – at the end of the Qing dynasty in 1911.

The new Republic of China replaced Classical Chinese with vernacular

Chinese as the writing system and vigorously promoted primary and secondary education. But an eight-year (1937-45) war with Japan and then a civil war – a total of 12 years of war – severely interrupted education and the growth of literacy. That is why the level was only 20 per cent at the foundation of the People's Republic in 1949.

Little in Zhou's early life prepared him to be "The Father of Pinyin". He was born in January 1906 into a cultivated family in Changzhou, Jiangsu province in east China; it was one of the richest and most advanced areas of the country. He was educated at modern schools in Changzhou and Suzhou and attended St. John's University in Shanghai; with English as its teaching medium, it was the most famous Western university in China at the time.

Zhou's specialty was economics. He went on to further studies at universities in Tokyo and Kyoto and learnt Japanese. He could also speak French. During World War Two, he worked for the Agricultural Bureau of the government, based in Chongqing, the wartime capital.

After 1945, he worked for Xinhua Bank in Shanghai, New York and London. Then, after 1949, he became a professor of economics at Fudan University in Shanghai. But he was always interested in linguistics. A prolific author with an encyclopaedic curiosity, he published articles and books on linguistics, as on other subjects that interested him. One of them, *The Subject of the Alphabets*, was published in Shanghai in

November 1954. It was read by Chairman Mao and other senior officials eager to reform the language. That is what led to the order to leave Shanghai and join the CCRC in Beijing in 1955. Zhou insisted he was a layman in linguistics and that economics was his field. But there was no refusing the order. It changed his life.

In 1982, the International Organization for Standardization (ISO) approved Pinyin as the global standard for written Chinese. This meant that it was used to write place and personal names, by post offices, aviation companies, immigration and customs officers and other official agencies. It was international recognition for Zhou's work. The United Nations adopted Pinyin in 1986. It is used by institutions and universities around the world.

In an interview in March 2009, Zhou said that one billion Chinese had used Pinyin to learn how to read and write. "It is not perfect but it has worked," he said with characteristic modesty.

His other great contribution came with digitalisation and the Internet. How do you write Chinese on a computer? While there are different systems, Pinyin is one of the most popular. Millions of Chinese use it, as does your modest author. About 900 million Chinese use the Internet and the number is rapidly catching up the 1.2 billion users of English. Dr Vinton Gray Cerf, one of the founders of the Internet, made a plaque in Zhou's memory. It reads: "His brilliant and persistent invention of Pinyin helped to bring the Internet and its applications within reach of the Chinese-speaking community. Long may he be remembered!"Outside Pinyin, Zhou had an extraordinary life. He lived through four Chinese dynasties – the Manchu, the Beiyang government, the Kuomintang and the Communists. Few people in the world live to the age of 111, and almost none retain to the end Zhou's intense curiosity and ability to write books and engage in serious dialogue with others.

He lived through the terrible campaigns of Maoist China, including the Cultural Revolution. In November 1969, at the age of 63, he was sent to a labour camp in the western desert region of Ningxia; he stayed there for 28 months. After the Cultural Revolution, he returned to work at the CCRC and retired in 1991, at the age of 85.

He was one of three Chinese editors of 10 volumes of *Encyclopaedia Britannica*, the first published in China. He lived in a small Beijing apartment, where he worked in a study of nine square metres. After his "retirement", he turned out books and articles on a wide range of topics. He retained his extraordinary curiosity and mental clarity and a wide circle of friends and admirers, Chinese and foreign. In total, he wrote 49 books. He died on January 14, 2017, one day after his 111th birthday.

Hundreds of millions of Chinese and foreigners, including this one, owe him a great debt of gratitude. Few scholars in history have changed the world as he did.

CHAPTER ONE

Blessed In Family And Education

Zhou Youguang was born on January 13, 1906 into a prominent family in Changzhou, Jiangsu province in East China. He was the sixth child of his parents and the first son; of the five girls, the first two had died young. His birth name was Zhou Yaoping; later in life, he changed it to Zhou Youguang. The Chinese empire was ruled by the Qing, Manchus from northeast China, who had founded the dynasty in 1644. In 1900, China had about 420 million people, accounting for a quarter of the world's population.

The young boy was blessed. He was born into a well-connected family with a comfortable standard of living, in one of the most prosperous regions of China. His family paid great attention to education, which would be one key to his success in life.

The family lived in a spacious home next to a river in the centre of Changzhou in the district of Qing Guo Xiang, an upmarket, residential area. As was common for wealthy Chinese, the family had servants to help take care of the children and run the house. Among the homes in the district, it stood out as it had been built during the Ming dynasty (1368-1644) and not the Qing.

Zhou's was a highly educated family; both his mother and grandmother were educated – rare among Chinese women at that time. The house had a large library. When he was just three, his grandmother taught him to read poems of the Tang dynasty (618-907). He spent much time with her. The love and erudition of his grandmother left a deep impression on him, the more so because one of the poems seemed to describe the scene of the river outside the window of his bedroom.

To this day, children in the Chinese world write and read these poems. My esteemed Hong Kong mother-in-law, 92, also copies them and recites them to us – a good way to refresh her memory and her knowledge of the characters. Zhou's parents invited private teachers to the house, to instruct his sisters in Chinese, English and dancing.

He had a distinguished ancestry. His great-grandfather was an official of the Qing dynasty (1644-1911) who later went into business in Changzhou. He owned a textile shop and a shop to sell its products, as well as a pawnshop. He became very wealthy.

But the family's fortunes were almost wiped out by the Taiping Rebellion of a peasant army calling for radical change and an end to the non-Han Chinese dynasty. It began in January 1851 in Guangxi, in the far southwest; the rebel army defeated the forces of the Qing government.

In September that year, it moved north and captured enormous areas of land. On March 19, 1853, it captured Nanjing; its leader Hong Xiuquan declared the city the Heavenly Capital of his kingdom. In May 1860, after defeating the Qing forces that had besieged Nanjing, the Taiping army moved toward the south of Jiangsu and Zhejiang, the wealthiest regions of the Empire.

Fearful for the safety of his family, Zhou's great-grandfather sent them out of the city; his daughters took what valuables they could carry. But he believed that it was his duty to resist the rebels whom he called the "long haired". He provided substantial funds to the small Qing army defending Changzhou. But it fell on May 26, and he committed suicide by throwing himself into the river.

In June 1864, Hong Xiuquan died and Nanjing fell to the Qing army. It was the bloodiest civil war in history, leaving over 20 million dead, soldiers and civilians. The war greatly weakened the Qing and left it increasingly less able to resist the relentless demands of the imperialist powers.

After the end of the rebellion, the Zhou family returned to Changzhou and resumed ownership of the land, where their factory, house and pawnshop had been. All the contents had been looted and there was no money. After the trauma of the rebellion, Zhou's grandfather became very conservative. He did not wish to become an official or be prominent in any way. He earned money by gradually selling off his assets. Zhou never met his grandfather.

In the south of Jiangsu province and the Yangtze Delta region, Changzhou lies in one of the most prosperous regions of China. After construction of the Grand Canal in 609, it became a port on the canal for the transshipment of grain. The counties around it produced rice, fish, tea, silk, bamboo and fruit; it became a flourishing commercial centre.

One result of this prosperity was excellent education.

The Changzhou Senior High School which Zhou attended had been established in a new form in November 1907, to provide modern education; this followed the abolition by the Qing government of the civil service examination educational system in 1905.

Changzhou is the birthplace of many famous people, including Sheng Xuanhuai, one of the most prominent entrepreneurs of the late Qing period, and Chao Yuanren, a famous linguist and later a model for Zhou in the language reform movement, was born in Tianjin, but his ancestors came from Changzhou.

Changzhou is 200 kilometres from Shanghai; the modernisation of which greatly affected the cities around it, including Changzhou, during the final decades of the Qing dynasty, Shanghai was the most advanced city in China, with industry, commerce, newspapers, publishing houses, schools and universities that could not be found in the rest of the country.

Most of the city was under foreign control – the International Concession, managed by the British and the Americans, and the French Concession, managed by the French. The writ of the Chinese government, be it the Qing or the Republic that succeeded it in 1911, did not run there. For Zhou and his classmates, this brought many benefits. The primary and secondary schools he attended in Changzhou offered a "modern" curriculum that included English, math and world history, as well as Chinese and Chinese history.

In primary school, he started to learn English. Only a fraction of schools across China offered such a curriculum. In his memoirs, Zhou greatly praised the quality of his teachers. Shanghai was also home to the most advanced schools and universities in China. One of them was St. John's, an American university which Zhou would enter in 1923.

Where is the Son?

Zhou's father was the only son of six children. During the Qing dynasty, the most important duty of a Chinese wife was to produce a male heir for her husband and his family. So it remains today in many parts of China, despite the efforts of different governments to promote the equality of male and female children.

A son maintained the family name and fortune and cared for his parents and grandparents in their old age; after the daughter married, she became a member of her husband's family. Her parents had "lost" her. Since

Zhou's father was her only son, Grandmother doted on him and pushed him in his studies. He was gifted; at an early age, he passed the exam for Xiu Cai, the first step on the ladder of the exam for the Imperial Civil Service.

The next stage was the Ju Ren, a rank achieved by passing the xiangshi exam in the imperial examination system of China. For this, he needed to take a boat trip to a nearby town, Jiangyin. But, en route, the boat capsized and he fell into the river. He found it so traumatic that he was unable to sit the exam. He returned to Changzhou and needed a long time to recover. He never took the exam again.

He spent the rest of his working life as a teacher, except for three years as the head of a township; the Chinese term for this is "Sesame and Green Bean Official", a vivid way to describe an official in a humble post. He earned his living from teaching and writing – but his income later proved to be not enough for a large and growing family with a middle-class lifestyle.

The family expenses increased after the birth of the fifth daughter. Fearful that her daughter-in-law would not provide the required male heir, Grandmother introduced a concubine into the house; she was a servant girl from a branch of the family in Hankou, Hubei province in central China. It was an immediate success – in the spring of the next year, 1904, she produced a son. Then, several months later, Madame Zhou produced her first son, the subject of our biography. Suddenly, the family had two sons.

The concubine went on to bear three more children. The expenses of raising all these children deepened the poverty of the family, as did Father's habit of smoking opium. Relations between Zhou's parents deteriorated. Then the young Zhou and his elder half-brother both contracted measles; at that time, it was a disease very hard to treat in China. The half-brother died and Zhou survived, leaving him as the only son.

On October 10, 1911, soldiers of an army in Wuchang, in central China, began an armed rebellion against the provincial government; it succeeded. It was the spark that lit a tinderbox of nationwide anger and discontent against a regime that was corrupt, inefficient and unable to reform.

Armies in other cities followed suit. In Nanjing, the revolutionary forces created a provisional government. On January 1, 1912, a National Assembly announced the establishment of the Republic of China, with Sun Yatsen as President. He was the leader of the Tongmenghui (United League), which he had founded with associates in Tokyo in August 1905; its aim was to overthrow the Qing government. It had set up branches in overseas Chinese communities around the world and underground ones in China itself. It launched rebellions in several cities in the mainland over the next six years; all were crushed by the Qing military, with heavy loss of life. By 1911, the Tongmenghui had almost run out of money. Sun himself learnt of the success of the Wuchang uprising in Denver, Colorado when he was reading a newspaper.

The end of the dynasty was greeted with jubilation by many in Changzhou, including Tu Yuanbo, principal of the Changzhou Senior High School, where Zhou would go to study in 1919. Tu was not only a school principal but also a revolutionary. He was an early member of the Tongmenghui; he served as the military commander in Changzhou during the Xinhai revolution.

An enlightened man, he had turned the school into a modern institution with a diverse curriculum. Each morning the students attended three classes. In the afternoon, they chose from electives arranged by the principal – classical literature, calligraphy, boxing, Chinese or Western music or playing in the school orchestra. There were no exams for these electives; you picked what interested you.

Tu was elected to the first National Assembly of the new Republic of China. So, in 1913, he moved to Beijing and resigned from the school. The new principal continued the policies of his predecessor. He invited famous people to give lectures to the students. Such principals were a blessing to their students.

Fluency in English

In 1912, Zhou started primary school near his house; his courses included Chinese, English and mathematics. He did so well that he graduated in six years, instead of the normal seven. At that time, written Chinese was Classical Chinese; if you wrote the vernacular language, that did not count. Zhou diligently studied the Classics and wrote compositions in the language. Then he moved to Changzhou Senior High School; it was an all-male boarding establishment, with students going home one day a week.

"The standard of English was very high," wrote Zhou. "World history, chemistry, geography and biology were all taught in English. So, when the graduates went to university, they could use English."

In later life, he knew the names of foreign countries and cities in English, not Chinese. This fluency in English would be essential to Zhou's success in life, enabling him to study at St. John's University, learn the economics and linguistics of foreign countries, and to work overseas.

With his classmates, Zhou imbibed the ideas of the New Culture movement that was sweeping China – modernisation, promotion of science and democracy, education for women and unbinding their feet, and replacing Classical Chinese with the vernacular as the written language.

But, while Zhou's studies prospered, life at home deteriorated. In 1918, his mother decided to move with her children to Suzhou. 80 kilometres to the southeast and closer to Shanghai. One reason for her decision was the presence of her husband's concubine. Another was the fact that, as a prominent family, the Zhous had many obligations in Changzhou. During Chinese New Year, the house was crowded from morning to night with relatives and friends; this meant financial obligations throughout the year – gifts for births, weddings, school graduations and anniversaries. The family could simply not afford them. A third was the fact that Zhou's father was an opium smoker.

While his mother and sisters moved to Suzhou, Zhou continued his studies at Changzhou Senior High School. Its rule was that, while local students went home for the weekend, those who lived further away stayed in the school. Zhou did not go to the home of his father, even though it was only a 10 minutes' walk away. Instead, he stayed on the premises and enjoyed the company of the other students who did not go home.

He joined his family in Suzhou only during the holidays. It was a long journey, on foot or by rickshaw, from Suzhou railway station to the family home. He became increasingly distant from his father. "My father smoked opium. This was the reason for the difficulties of the family. When visitors came, he invited them to smoke opium or smoke cigarettes. The lifestyle of the old families was that everyone smoked opium. My mother saw that such a large family could not continue such spending and so decided to move to Suzhou."

Zhou's father rarely went to Suzhou. He and his wife grew increasingly distant. He gave her a small allowance, but it was not enough to cover the living costs of a household of six people. Mother and her children became increasingly poor; they had no other source of income. Their situation improved when one sister found a teaching job in Shanghai. Another went to teach in Southeast Asia; she founded a ladies' Overseas Chinese Middle School in Rangoon, the capital of Burma. Both daughters sent money to Mother; her financial situation improved, but only enough to provide a minimum living for the family. In Suzhou, they lived in rented accommodation, often not ideal.

In 1923, Zhou graduated from Changzhou Senior High School. After his good results at the school, he wanted to continue his studies at university. Most prestigious was St. John's University in Shanghai. It had been founded as St. John's College by two American Anglican missionaries in 1879 and began with 39 students, teaching mainly in Chinese. It was one of 14 universities established by foreign missionaries in China. In 1881, it switched to teaching in English, becoming the first college in China to do so. Only Chinese language and Chinese history were taught in Mandarin. It included science and natural philosophy in its courses.

In 1905, it became St. John's University and registered in Washington D.C.; this gave it the status of a domestic university and its graduates could proceed directly to graduate schools in the U.S. Many called it the "Harvard of China". As a result, it attracted some of the brightest and wealthiest students in Shanghai. It set up colleges of literature, science, medicine and theology. From 1907, it was the first institution to grant bachelor degrees. It was located at 188 Jessfield Road (now, Wanhangdu

Lu), on a bend of the Suzhou Creek in the International Concession in Shanghai; its design included Chinese and Western architecture. Its graduates included many of the most famous Chinese of the 20th century, including Wellington Koo, the country's representative at the Versailles Peace Talks in 1919 and later Foreign Minister: architect I.M.Pei and author Lin Yutang.

To enter St. John's meant passing an entrance exam. This ran for six days, with two papers of three hours in the morning and afternoon. More than 80 per cent of the papers had to be answered in English. Zhou needed an overall score of 70 per cent to pass. It was a tribute to the quality of teaching, especially of English, and his diligence as a student that Zhou did not find the exam difficult and passed comfortably. He applied to one other university, the Southeast Normal College in Nanjing; it later became Central University. This was one of six colleges set up after the Xinhai Revolution to train secondary school teachers. To enter, Zhou had to sit another exam; he also passed it comfortably.

Of the two, Zhou preferred St. John's – but the fees were a big obstacle. The Nanjing college charged no fee and gave students a subsidy. St. John's, a private institution, charged 200 silver dollars per term. It was the moment of greatest poverty of the Zhou family; they could not possibly pay such a large sum. So the young man was preparing to go to Nanjing. Then, out of the blue, he was saved by an act of great generosity. His sister, a teacher in Shanghai, mentioned his situation to a colleague at work; the colleague said it would be a great pity for Zhou to miss the opportunity to enter such a famous university simply for lack of money. She asked her mother to help. Her mother lived in Suzhou and very much liked the young Zhou; she was willing to help. She said that, while she did not have cash, she had a chest full of dowry gifts she had received on her wedding. The chest had been sleeping in an empty room in her house for many years; its contents were no use to her. She took the chest to a pawnbroker and borrowed the money Zhou needed to go to St. John's.

But for her generosity, Zhou would probably have become a secondary school teacher and the world today would never have heard of him.

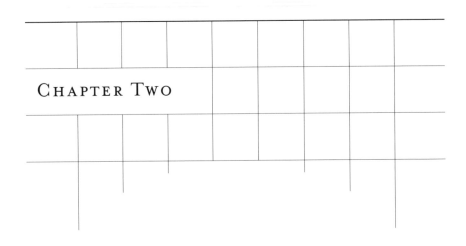

Two Universities, A 'Profitable' Marriage And Missing Professor In Japan

In the autumn of 1923, Zhou took a train from Suzhou to Shanghai station; then he took a tram to Jing An Temple and a rickshaw to his new home, St. John's University. Then 44 years old, it was one of the most modern and most Westernised universities in China. It had a range of facilities few universities in the country could match – a Science Hall built in 1894, the first in China dedicated to teaching natural sciences.

By 1905, it had five colleges – arts, science, medicine, engineering and theology. That year it was registered as a university in Washington D.C.; this meant that its graduates could go directly to the graduate schools of American universities.

By the 1907-08 academic year, 30 of its graduates were studying in the U.S. and 10 in Britain. A second Science Hall was built in 1918 with state-of-the-art laboratories and, biology, chemistry and physics departments. In 1919, it opened a gymnasium, with an indoor swimming pool and basketball courts.

It had tennis courts and a large playground where people could cycle. It paid for these facilities through student fees and donations from alumni, a system successfully introduced from the United States. This enabled the university to develop on its own; this spared it from China's chaotic politics and the erratic funding that plagued state universities for the first 30 years of the 20th century. It had a large, leafy campus with many trees. Since 1881, it had been teaching in English; the majority of the faculty were foreigners. Many called it "the Harvard of China".

On arrival, the first thing Zhou did was to register. Staff gave him a name card, with Chinese and romanised spelling using Shanghai dialect. "The school management was done according to this romanised spelling. This was scientific management, it was excellent. Even until today, Chinese do not understand," Zhou said in an interview in 2010. It was the first time he had seen romanised spelling for Chinese characters. It aroused a curiosity that would last 80 years.

At St. John's, Zhou chose economics as his major. For a second foreign language, he chose French, taught by an elderly lady who brought her dog to class. The university had an excellent library, which he used frequently; there he read *Das Kapital* of Karl Marx translated into English. He also attended lectures on linguistics as an elective, because the subject interested him.

He fitted seamlessly into his new surroundings. Thanks to the high-quality education of his two schools in Changzhou, his English was good enough to understand the professors comfortably. He appreciated the methodical Western style of teaching, on subjects like international trade.

"Many of the teachers were foreigners," he said. "They were good. They encouraged us to ask questions and develop a critical faculty. Compulsory lessons were not so many. The teachers wanted to develop our interest in subjects and pursue them, not just for exams. It was a Western way, not the Soviet way we adopted after 1949 of compulsory and 'stuffed-duck' education."

He developed an interest in the romanisation of the Chinese language. From his primary school age, he had learnt English and seen how much easier it was to use than Chinese. The vast majority of his compatriots were illiterate and could not read characters; how could they learn? His idea was not to abolish the characters but to develop a system of Pinyin in Roman letters that could be used on a typewriter. So, he was delighted that such a system was being developed. While he did not have the money to buy a typewriter, some of his classmates did and he borrowed theirs. It was the start of a mission to which he would devote much of his life.

Among his fellow students at St. John's was Song Zian, a brother of the three famous Song sisters. One married Dr Sun Yatsen, one married President Chiang Kaishek and one married Kung Hsianghsi, Minister of Finance from 1933-44 and Governor of the Central Bank of China from 1933-45. There was a Chinese joke about the sisters that said: "One loved money, one loved power and one loved the country". On Saturdays and Sundays, two of the sisters drove a car to St John's and took their brother away for outings. "We often saw Song Chingling and Song Meiling," Zhou recalled. "We did not imagine Meiling would become the wife of Chiang Kaishek."

The Song family belonged to China's small, wealthy elite; Zhou was a poor, if talented student. In 1926, Song Zian went to Harvard and obtained a Masters in Economics two years later. After he returned to China, he held senior jobs in banks and finance but did not go into politics, like many of his family. Before the Communist revolution of 1949, he moved to California and divided the rest of his life between there and Hong Kong, where he died in 1969. After 1949, the St. John's campus was taken over by the new government. It became part of the East China University of Political Science and Law. Those who wish to see the buildings of the "Harvard of China" can go to its campus in the Changning district of Shanghai.

In Protest, Move to New University

Despite his affection for the teachers and educational system of St. John's, Zhou left the university after two years, halfway through his course. This was a result of the political turmoil sweeping Shanghai. It was the most modern and developed city in China, with thousands of factories, many foreign-owned, and a modern media that did not exist in the rest of the country.

As a result, the workers were better informed and more unionised than in other cities. There were disputes over wages and working conditions. Added to this were the toxic ingredients of "nationalism" and "imperialism". China's poor treatment by the victorious powers at the Versailles Peace Talks (1919-20) after World War One had unleashed a wave of anger against the colonial powers, especially Japan. This was stoked by the two major political parties who were competing for public support – the Nationalists and the Communists, who had established their party in Shanghai in 1921. It meant industrial disputes with foreign owners had a bitter nationalist edge.

The largest cotton manufacturer in Shanghai was the Japanese-owned Naigai Wata Company. All its nine mills fronted on Soochow Creek. Close by the mills were mess halls and dormitories for workers, company nurseries and infirmaries, spacious residential quarters for Japanese employees and a park for the use of the staff. Its Number Eight Mill was a place of conflict between the owners and the Chinese workers. In February 1925, a group of Japanese managers was attacked while leaving work and one was killed. As a result, the Japanese foremen started to carry pistols on duty. On May 15, a group of workers forced their way into the plant, which the management had closed due to a shortage of raw materials. The workers were armed with sticks, iron bars and other weapons. Outnumbered, the Japanese opened fire and wounded several, including one named Gu Zhenghong; four Japanese were seriously wounded in the clash. Two days later Gu died in hospital. He became a national martyr.

On May 30, the situation worsened when a crowd of 1,500-2,000 Chinese gathered outside a police station to demand the release of leaders of a student protest arrested that morning. The crowd became increasingly violent and was about to enter the station; the police – Sikhs and Chinese led by three European officers – opened fire, killing four demonstrators; five others later died of their injuries.

This violence provoked widespread protests across Shanghai, especially among students. Many at St. John's University wished to join. But the foreign administrators who ran the university would not allow them to leave the campus. Liberal as they were, they told the students that they could hold meetings on campus to discuss the issues. They feared for the safety of the students once they left the campus and joined emotional and violent protests outside.

The city was controlled by British, American and French authorities who favoured the business community; their police regarded the protesters as troublemakers and a threat to social order. Add to this dangerous mix the agitation of the Nationalist and Communist parties. The anxiety of the university chiefs for their students was well founded. But such was the depth of emotion among the St. John's faculty and students that, as the price to pay for joining the protests, they were willing to leave this prestigious university and give up their future degrees.

On June 3, 553 students – Zhou among them – and all 19 Chinese teachers of St. John's announced that they were leaving the university; more than 10 graduates said that they would not accept its diploma, a decision that would gravely affect their future. Their resignation made headlines around the world. The very next day, they issued an appeal for land and funds to build a new university and evoked the patriotism of fellow Chinese to help.

Such was the response, especially from overseas Chinese, that, within three months, Kwang Hua University was established. This was no simple matter. The government had no spare money to fund such an institution, so it was entirely a private enterprise. The father of one of Zhou's classmates, a large landowner, donated 100 mu (one mu is equivalent to 0.165 acre) of land to build the campus. Then a man who held senior positions in the finance departments of Jiangsu province and Shanghai city raised a substantial sum of money.

Not all the students who left St. John's were willing to wait to join it; some went to other universities, where they were welcomed. Others returned to their hometowns. As a new college, Kwang Hua had no history and no prestige; what would be its future?

Zhou himself returned to Suzhou and asked his mother what to do: take a job to relieve the poverty of his family or continue his studies? She left the decision up to him; she said that she had no money to help him. He was on the horns of a dilemma. "If I found a job, my life would be more stable and I would not have to worry so much about money. But I feared that I would have no future without more study. It was a crossroads in my life. I felt very unhappy and took no decision."

Without money or connections among the country's elite, Zhou had no hope of rising in society without academic qualifications; two years and no degree from a university, even a famous one, would not count for much. At that moment, Zhou received a letter from one of those setting up Kwang Hua; he asked Zhou to return to Shanghai to join the preparation work. The work would not be paid but, once the new university opened, they would be able to settle the issue of his fees quickly. Zhou accepted the offer.

The organisers rented space in Avenue Joffre (now Xia Fei Lu) in the French Concession. The 19 Chinese teachers from St. John's formed the core of the faculty; they were later joined by other outstanding professors. Zhou was able to resume his studies in the new institution. The project sparked a wave of patriotism not only among people within the country but also overseas Chinese; many of Zhou's fellow students were overseas Chinese. Their families were very generous, donating large sums to the project. Land in Shanghai was the most expensive in China.

Zhou said that, given all the difficulties, the project went quickly and smoothly and the new university opened quickly. It was a tribute to the generosity and patriotism of the donors and organisers. It attracted many famous teachers, including Hu Shih, an esteemed intellectual who advocated language reform. Many gave their classes in English. So, for Zhou and the other students, the high quality of teaching continued from St. John's to the new campus.

Since Kwang Hua was, like St. John's, a private university, it charged fees of at least 100 silver dollars a year. For Zhou, this was a problem. He overcame this with the help of the university principal who gave him teaching work at a middle school attached to Kwang Hua, as well as administrative work in his office. These duties covered part of the fees.

In 1927, Zhou graduated from Kwang Hua with a B.A. in Political Science. He seems to have considered the four years as one stretch. In Kwang Hua, he studied romanisation, part of the debate on what to do with Mandarin. Should they romanise Beijing dialect or Shanghai dialect? This raised many issues that did not come up with the study of English. It was the start of his interest in linguistics.

After graduation, most of his classmates continued their studies, in China or overseas; they were children of wealthy families. But Zhou did not have the money. The principal continued to help him, giving him teaching work in the middle school and in the university itself. After he started work, he felt unwell and consulted a doctor, leading to the discovery of early-stage tuberculosis. Zhou was fortunate to find it early. "I took cod liver oil for one year and recovered. This treatment was light."

Romance Blooms

Zhou found his girlfriend thanks to an introduction by his younger sister; she was a student at a private school, Suzhou Leyi Ladies' Middle School and a classmate of Zhang Yunhe. The two girls became friends and Miss Zhou often brought Miss Zhang home where she met the elder brother.

She came from a remarkable family. The school the two girls attended had been established by her father who had provided the funds and the buildings. Her great-grandfather, Zhang Shusheng had been one of the most important officials of the Qing dynasty; he was a governor of Jiangsu, Jiangxi, Guangdong, Guangxi provinces. He had negotiated with France, Holland and Japan over Vietnam, Taiwan and Korea respectively.

His son, Zhang Jiyou, father of Yunhe, was also an official but not as senior as Shusheng. The family was wealthy, and very enlightened. It had a deep interest in Kunqu, one of the leading schools of Chinese traditional opera; it became a lifetime passion of Yunhe and her young sister Zhang Chonghe. Zhang Jiyou was a friend of many of China's leading reformers, including Cai Yuanpei and Jiang Menglin. Jiyou did not want to leave his money to his children, the norm in Chinese society then as today. So he invested the money into two middle schools in Suzhou, one for boys and one for girls; his daughter and Zhou's sister went to classes there. The money for the first ran out and it closed. But the girls' school was smaller and remained open until 1949, when the new government took over all the private schools.

Zhou and Yunhe liked each other; but the gulf between the wealth of the two families was very wide. The Zhou home in Suzhou was small and bare. Zhou was shy and serious; Yunhe was self-confident and talkative. "I said that I was very poor and feared that I could not bring her happiness. She wrote back a 10-page letter, saying that happiness was something we created by ourselves."

She went to study at a middle school in Shanghai, while Zhou was at Kwang Hua University. Her school was in Wusong, a suburb to the north of the city on the banks of the Huangpu river. On weekends, he made the long journey there and took her out to lunch. The two walked along the river, next to the large stones that acted as flood protection.

The two exchanged letters and gifts; there were no telephones or Internet in those days. They gradually fell in love. "She liked Classical Chinese music and I liked Western music," he recalled. "In 1927 or 1928, I invited her to a concert of Beethoven in the French Concession. Each ticket cost two silver dollars, very expensive. She slept through it. It was a joke. She did not like Western music as much as I, nor did I like Chinese music as much as she. After we married, we went to concerts of both with the other." Neither ever fell asleep again. She especially liked Kunqu. After graduating from her secondary school, she went on to study at Kwang Hua University.

Teaching Workers and Farmers

In 1928, Zhou moved from the comfort of living and working in Shanghai to a rural college in the outskirts of Wuxi, an industrial and textile city

In 1930, Zhang Yunhe was the president of the student union of Kwang Hua University.

135 kilometres to the west. Who persuaded him to resign from Kwang Hua University? It was a friend and fellow professor named Meng Xiancheng, 12 years his senior. Also a native of Changzhou, Meng had also studied at St. John's University; he received a government scholarship to study at George Washington University in Washington D.C. and later the University of London. After returning to China in 1921, he taught at St. John's and was one of the faculty who left to join Kwang Hua.

Zhou described how Meng persuaded his young friend to follow him to Wuxi. "'Working at Kwang Hua is of course very stable. But I advise you to go somewhere else to work and not always stay in the same place.' Meng understood Western thinking. A person should not stay very long in one place, but should change his job and enrich his experience. He was going to set up a new college in a rural area of Wuxi, aimed at educating ordinary people. There was a new theory in capitalist countries that education should be for the general public, not the elite. So, I went with him."

One of Meng's inspirations was a Danish pastor and teacher named Nikolaj Frederik Severin Grundtvig, one of the most influential people in Danish history. He advocated education for the general public, not for the elite; schools should train people for active participation in society and the life of ordinary people.

For the young Zhou, his new environment was a challenging experience; students at the rural college were not the well-dressed and well-educated young people among whom Zhou had spent his life. They were workers and farmers, with a poor level of education and literacy. He threw himself eagerly into the work and helped to edit a monthly magazine *Education and Common People*; he wrote articles introducing adult education in England, Poland and Estonia. One downside of moving to Wuxi was that his romance with Miss Zhang cooled; with her in Shanghai, they saw less of each other.

In 1930, the Zhejiang provincial government appointed Meng head of a similar college for popular education in Hangzhou, the provincial capital. Meng took Zhou with him and put him in charge of teaching English and Western history. His years in Wuxi and Hangzhou gave him a regular salary. It greatly enriched his experience outside the comfortable lecture rooms, libraries and campuses of Changzhou and Shanghai. He met hundreds of people who were semi-literate or could not read at all; it made him reflect on the nature of the Chinese language, its multiple dialects – often mutually incomprehensible – and what its written language should be.

On January 2, 1931, the Zhang sisters held a cultural soiree at the family's large home in Suzhou; Zhou's girlfriend invited him to attend. He met her parents for the first time. He recalled later: "Her father and mother were very nice to me. By the standards of that time, her father was very open-minded, proposing free love to his daughters. Many people used the old method to court their daughters. Father said: 'Marriage is for them to decide freely. We do not interfere.'" After this public approval, Zhou and Miss Zhang could be more public in their romance; her sisters accepted him as their future brother-in-law. This was not the norm in Chinese society at that time; it was families who arranged marriages for their children.

The move to Hangzhou turned out to have an unexpected benefit. At the end of January 1932, the Japanese military attacked the Chinese army in the Zhabei district of north Shanghai. Fierce fighting spread to the Wusong district, where Miss Zhang was attending university. She and her fellow students fled. With the railway line cut, they took a boat to Suzhou.

Since she could no longer continue studying in Shanghai, she moved to Zhijiang University in Hangzhou; it was an institution established – like St. John's – by American missionaries.

In Hangzhou, she and Zhou saw much of each other; their relations deepened. He later said that their relationship had three stages. "The first stage was in Suzhou, when everyone went out together. The second was in Shanghai, when the two of us went more together. The third stage was in Hangzhou. The scenery in Hangzhou is beautiful and the best place for love. This was the stage of love.

"In Hangzhou, we went out on Sundays but did not hold hands. One day a monk followed us, walking slowly or quickly as we did. He purposefully listened to what we were saying. When we sat on the top of a large stone, he sat next to us and asked me: 'How long has this foreigner lived in China?' I said that it was three years. 'No wonder her Chinese is so good!' My wife's nose was a little larger than that of most people. The monk took her for a foreigner!"

One of the prettiest cities in China, Hangzhou provided many walks and scenic spots for the romance to flourish; best was the West Lake, where they rented a boat and explored its nooks and crannies. Originally, Miss Zhang planned to return to Shanghai to continue her studies. But the war there made it impossible, so she graduated from the university in Hangzhou.

Meeting Japanese in Manchuria

During this period, Zhou and his mother made a visit to Manchuria in northeast China. They went to Changchun. In September 1931, the Japanese military had invaded the three provinces of northeast China and turned them into the puppet state of Manchukuo; it was formally established on March 1, 1932. It chose Changchun as the capital of its new state and renamed it Shinkyo, which means "new capital".

One of Zhou's elder sisters had gone to Japan to study; her husband had studied there for a long time and specialised in medical products. He had taken a job in a department making drugs attached to a railway hospital in Changchun. The railway system of Manchukuo was the most developed in China. Zhou's sister and her husband invited Zhou and his mother to come and stay with them; they paid all the expenses, including the train ticket.

"This trip to the Northeast was very interesting to me, not as tourism but to see how to develop the economy of China. I saw the rich mineral resources of the Northeast and the forests, not yet developed. Not like today, when the trees have been cut down," said Zhou later.

For the first time in his life, he saw many Japanese in Changchun, where they had opened shops and restaurants. The winter was cold and the rivers froze over. Zhou and his family went skating on the ice and caught fish after drilling holes in the ice and dropping a line down. His mother and sister were eager to arrange his marriage with Miss Zhang. Fortunately, as we described earlier, her parents had given their consent.

The couple did not want the extravagant, expensive wedding chosen by many wealthy families. So, they selected a very modest option – the YMCA in Shanghai – as the venue. They chose a date and printed 200 invitations; but a grandaunt objected, saying that the day was inauspicious. That meant they had to choose another day, April 30, 1933; they tore up the invitations and printed new ones. Zhou paid 20 silver dollars for a new suit, compared to 50 for one from a high-class tailor; his bride rented a wedding outfit for the day.

They held the ceremony in the meeting hall of the YMCA, with space for more than 100 people. It was the simplest and least expensive they could have in the city. Among those present was Zhou's mother and Zhang's father and second wife; her mother had died when she was 12. Zhou's father did not attend – a reflection of the bad relations between the two.

The music was provided by Ora, a 14-year-old White Russian living in Shanghai and an accomplished pianist; and one of Miss Zhang's sisters sang a piece of Kunqu opera, accompanied by a bamboo flute. The guests, 102 in number, ate Western food in the hall, with a portion for each person. Most of the guests were members of the two families.

The custom was for guests to give money on arrival. Most gave two silver dollars. Those from the Zhang family were wealthier and gave more. Some gave gift coupons from the big Shanghai department stores; you could use them to buy items, exchange for money at the stores, or even put the money into your bank account. After the wedding, the couple discovered to their delight that they had made a "profit" – they had received 800 silver dollars in donations, double the cost of the event! They put the difference into their bank account to fund future studies. In addition, Miss Zhang's parents made a gift of 2,000 silver dollars, a substantial amount at that time. This too they put in the bank.

Given that she came from such a prominent family, they could scarcely have had a more modest marriage. A wedding was an opportunity for the rich to show off their wealth, in China as in other countries. A male cousin of Miss Zhang spent 5,000 silver dollars on his wedding.

The Zhang family had four daughters and six sons. The third daughter was

Zhang Zhaohe; of the 10 children, she had the most famous marriage. She was studying at a renowned secondary school in Shanghai, China Public School and had many suitors. Among her teachers was a well-known novelist, Shen Congwen. He fell in love with her and wrote many letters. Some she read, others she did not; but she kept all of them. Frustrated at her lack of response, Shen asked the school principal Hu Shih to intercede for him. Hu told Miss Zhang that Shen was very much in love with her. "Yes," she replied. "But I am not in love with him." Despite this setback, Shen continued his pursuit by letter and finally won the heart of his beloved. He went to Suzhou to seek the approval of her parents; they said that the decision was up to their daughter. So, the couple married in 1933 and moved to Beijing; they would have two sons.

Shen went on to become one of China's most famous novelists in the 20th century. Between 1933 and 1937, he produced more than 20 volumes of prose and fiction. After 1949, he stopped writing fiction but wrote books on calligraphy, art history and other aspects of Chinese art and culture. In 1987 and 1988, he was nominated for the Nobel Prize in Literature; in 1988, he made the list of finalists but died before the prize could be awarded to him.

Zhou and his new wife spent the first night of their married life in a Shanghai hotel. The next day they returned to Suzhou. They wanted to go abroad to study. The natural choice of a student of St. John's was the United States. But their savings, even with the generous gift of Miss Zhang's parents, were not enough to pay for study there; and they did not know if they could work there to fund their studies. "My sister and her husband had studied in Japan and thought Japanese would be very useful," Zhou said. "If a person had learnt English, he could learn Japanese. I had studied it myself and could read books and translate. But my listening ability was poor. My sister and brother-in-law advised us to go to Japan,

Zhang Jiyou and his daughters, Zhang Yunhe, Zhang Chonghe, Zhang Zhaohe, Zhang Yuanhe (from right to left) at Jiuru Lane in Suzhou.

where fees were cheaper. It was close, so that, if anything happened, we could come back. First study in Japan for several years and then go to the U.S. We agreed, saying that going to the U.S. was too great a risk. There was no risk in going to Japan, so we agreed."

Japan – the Missing Marxist Professor

One cold October evening in 1933, the couple boarded a ship in Shanghai for Japan. The next day they reached Kobe and then Tokyo, where they stayed in the Chinese YMCA. Since Miss Zhang had not studied Japanese before, she enrolled in a Japanese-language class. Since Zhou already had a certain fluency, he did not attend the classes. His first choice was Kyoto Imperial University. He wanted to study under Hajime Kawakami, a Marxist professor of economics, whose books he had read; his translated works were popular and influential among Chinese students.

A graduate of Tokyo Imperial University, Kawakami had earned a professorship at Kyoto Imperial University. But he became increasingly leftist and was expelled from the university as a subversive. He joined the outlawed Japanese Communist Party, was arrested in 1933 and sent to prison for four years. After his release in 1937, he translated Karl Marx's *Das Kapital* from German to Japanese. But Zhou knew nothing of this. He thought that the professor was still in Kyoto. To take the entrance exam, he travelled to Kyoto. To his surprise, he discovered that Professor Kawakami no longer taught there and was under arrest. "I did not know about this, I was ill-informed," he said. "But, never mind, Kyoto University had this tradition, so I took part in the exam anyway."

The exam lasted three days; because his written Japanese was not good enough, he answered the questions in English – he knew the university would not accept answers in Chinese. To his surprise and that of his friends in Tokyo, Kyoto Imperial University accepted him. He also discovered that, since Japan did not recognise Chinese degrees, it would take him three years to obtain a Ph.D. – against one-two years in the U.S., which did recognise them. "I did not want to study again what I had already learnt. So, I had no specialty. I mainly studied Japanese language, as well as Japanese culture and lifestyle. I did not waste my time. A good command of Japanese was very useful. I used it when I later went to the United States."

His wife became pregnant; in early 1934, she decided to return to Shanghai and live with her family. Material conditions and family support there would be better during her pregnancy than as a student in Japan. In addition, she did not find the country interesting; her language ability was poorer than that of her husband. But he wished to stay; so, she returned to Shanghai on her own.

Zhou remained in Tokyo for a short while and then moved to his studies in Kyoto. He found it a great contrast to the capital – quiet and with a small population. He took a room in the home of an elderly lady who rented rooms to Japanese and foreign students. She was very kind to Zhou and taught him features of Japanese life different from those in China, such as how to take a bath and to wash your face in the morning with water mixed with ground soya bean. To his surprise, Zhou found the dialect of Japanese spoken in Kyoto different to that of Tokyo, which was, and is, the standard language; he had to adapt.

One rule at the university was that, before students started their classes, they bought the course material from designated shops; it was delicately printed on good paper. This enabled Zhou to study a great deal on his own. This became a habit he would maintain throughout his life; it was one reason for his very long life – the ability to absorb himself in subjects that interested him in his small study in a Beijing lane, without the need for a teacher or colleagues. "The study atmosphere of Kyoto Imperial University at that time was very free. You could visit a professor and ask him questions, which he answered. That was very good. But, while studying was very relaxed, the exams were very strict."

On April 30, 1934, their first child was born, Zhou Xiaoping (周小平). When he was a student, he later changed it to Zhou Xiaoping (周曉平). In 1935, Zhou decided to return to Shanghai and rejoin his wife and baby. The son was followed by a daughter, Zhou Xiaohe. Zhou's wife did not wish to return to Japan; she did not find it interesting. Friends looked at Zhou's language and social skills and suggested he apply to the Foreign Ministry to work as a diplomat. His wife intensely opposed this. He said later: "I think she was right. If I had gone into politics, it would have meant trouble."

His aim after his return to China was to work for a short time and then go to study in the United States. His wife was teaching at an experimental Middle School under Kwang Hua University and he joined her there. "On one side, we were teaching. On the other hand, we were preparing to go abroad to study. But I had no network of contacts in the U.S., unlike Japan, when I knew many people. In addition, the university fees in the U.S. were high."

What does this period of 18 months of study in Japan tell us about Zhou? First, it shows his energy and intense intellectual curiosity to make friends, study a new language, a new culture and a new society and do all this well enough to pass a difficult university entry exam. Even after his wife returned to Shanghai, he was ready to study on his own. Second, it shows his willingness to make friends with Japanese, including professors, fellow students and ordinary people, despite the fact that this was a period of extreme tension between China and Japan. His country was already at war with Japan after its Guandong Army occupied the three provinces of China in 1931 and declared the new state of Manchukuo.

The Foreign Ministry in Tokyo had ordered the Guandong Army to stop military operations after the capture of Shenyang in September 1931, but it ignored the order and had taken over the whole of Northeast China. On May 15, 1932, naval officers assassinated Prime Minister Tsuyoshi Inukai.

In 1933, Japan walked out of the League of Nations because of its refusal to accept the conquest of Manchuria. The military was increasingly assertive and violent; it aimed to overthrow the party-based political system put in place after the Meiji Restoration of 1868. The military presented Japan as the master race of East Asia and China as a weak child; Japanese soldiers wrote slogans such as "Slaves of a Dead Nation" in buildings they had occupied in China. Many citizens of each country loathed those of the other. Despite this, Zhou was able to find the space and manners to continue his studies. He lived in a world of professors, students and people who treated him with courtesy and appreciation. He was able to concentrate and have constructive relations with Japanese despite this surrounding poison of nationalism and xenophobia.

CHAPTER THREE

Becoming A Banker, Fleeing Down The Yangtze

"In life, things often work out in ways you do not expect." Zhou aptly summarised what happened after his return to Shanghai in 1935. He came back to China with a sense of mission incomplete; his period of study in Japan had given him an understanding of its language, culture and society. But its universities were not linked to those in China and he did not obtain the Ph.D. he hoped for.

His conclusion was that Chinese should go to the United States, not Japan, for further study. He wanted to follow the example of hundreds of talented Chinese since the final years of the Qing dynasty 40 years earlier. After they graduated from American universities, a few chose to stay in the U.S. The majority returned to China and, blessed with skills and knowledge rare at home, found important posts in the government, universities and business. They joined the elite of the new Republic.

But destiny had other plans for Zhou. In Shanghai, he met a university classmate who had landed an excellent job in the Bank of Jiangsu, one of eight major banks in the city. It was founded in Suzhou on December 3, 1911, making it the first local bank established after the overthrow of the Qing dynasty on October 10 that year; the founder was Chen Kwangpu.

In February 1912, Chen moved the company's head office to Shanghai, China's financial and economic capital. Chen had started his career in Hankou in the import-export business where he had learnt English. He won a Boxer Indemnity Scholarship in the United States, earning a BSc degree from the Wharton School at the University of Pennsylvania in Philadelphia, before returning to China in 1910. The scholarships came about when the U.S. decided to give back to China some of the money it received in compensation for the Boxer Rebellion (1899-1901), using it to pay for Chinese to study in the U.S.

Chen then created a bank on the Western model, run by a board of directors, employing foreign accountants and publishing its accounts once a year. He encouraged staff to use English as well as Chinese. The government of Jiangsu province provided some of the bank's capital, but the majority came from private shareholders. This meant that the board decided where and how to lend the bank's money – it was not an arm of the Jiangsu government.

Zhou's friend was personal secretary and advisor to the managing director of the Bank of Jiangsu. He told Zhou that the bank needed young people like him trained in modern finance and economics. Many of the staff were middle-aged men who had worked in traditional Chinese banks, before the introduction of modern Western methods. He urged Zhou to accept a job in the bank. Zhou explained his disappointment over his time in Japan and his wish to continue studies in the U.S. In response, his friend said the U.S. would always remain an option but that the job on offer was a rare opportunity that would not come again; he urged him to take it while it was available.

The Bank of Jiangsu was an important institution, with a licence to print money. At the same time, Zhou met a Kwang Hua classmate who had become a professor at Kwang Hua University. He invited Zhou to go and teach there.

Zhou was attracted by both offers. He discussed them with his family. Their consensus was that the job in the bank was too good to turn down. It offered a stable position and excellent remuneration. So, he accepted it; he also won the agreement of the bank to teach part-time at Kwang Hua. This gave him what he considered the ideal combination of work; it also meant long hours and an exhausting routine. With such a schedule, his wife, son and daughter chose to stay in Suzhou, together with his mother; his wife had resigned from her teaching job after the birth of their daughter.

Zhou's handsome new salary from the bank gave them the money to rent a large house in Suzhou from a family whose children were all living overseas. The family preferred Suzhou as quieter and cheaper than Shanghai, especially for the children. The house, on the southern edge of the city, looked out over a beautiful rose garden; it was quiet and peaceful. Usually Zhou went there on Saturdays, stayed the night and returned to Shanghai on Sunday afternoons. He lived in a small room on the fifth floor of the Jiangsu Bank building.

When he stayed in Shanghai for the weekends, he sometimes took his wife to a dance hall or the cinema, or had meals or coffee with his friends. Almost all the films were American and silent, without subtitles in Chinese. The dialogue was written in English, at the bottom of the screen; graduates of secondary school or university could follow it.

One of their favourites was *Seventh Heaven*, a 1927 romantic drama about a young man named Chico who works in the sewers of Paris. One day he saves a young prostitute and, to keep her from arrest by the police, pretends to be her husband. She moves into his flat and they fall in love. After World War One breaks out, she goes to work in an ammunition factory and he is sent to the front. She is told of his death – but he returns, wounded and blind. Other favourites were *Romeo & Juliet* and *Faust* by Johann Goethe. Since the films were silent, Filipino bands provided music – but, when sound was later added, they were no longer needed and had to return home.

Zhou's work at Jiangsu Bank enabled him to meet those working in other large banks in Shanghai in associations which they managed jointly. He expanded his network of friends and contacts outside the academic field.

As if Zhou was not busy enough with the two jobs, he also joined an organisation preparing the population for the war with Japan that everyone expected – the Save the Nation Society (SNS). It set up committees in many sectors of life, including banking; Zhou joined this committee. This meant additional work, usually in the evenings and at night; much was secret, to keep information away from the ears of the Nationalist government. For them, the priority was to wipe out the Communist Party and then deal with the Japanese.

Zhou wrote articles and speeches for the SNS newspaper, many aimed at students. In December 1936, Zhang Xueliang, a general from Manchuria, kidnapped President Chiang Kaishek in the central city of Xi'an; he wanted to force him to agree to joint military operations with the Communists against Japan.

"After the Xi'an incident, I was very nervous," Zhou said in his oral autobiography. "I believed there were two possibilities. One was that the Nationalists and Communists would intensify their war. The Nationalists would put up another leader and sacrifice Chiang Kaishek ... the other possibility was that the Nationalists and the Communists would co-operate."

Individuals could not control nor predict the relations between the two parties nor those between China and Japan. "The situation of individuals was all put into second place," he wrote. Finally, Chiang agreed to fight with the Communist army against Japan. The civil war was put on hold, as it turned out, for nine years.

Looking back in later life, Zhou remembered this period of more than two years in Shanghai before the Japanese war as one of comparative tranquillity. "The eight years of war changed everything. They changed our feelings," Zhou said.

Escape to the Southwest

At 15:00 on August 13, 1937, Japanese forces crossed a bridge in the north of Shanghai and attacked Chinese troops. It was the start of the Battle of Shanghai; it was one of the largest engagements of World War Two and involved nearly one million soldiers on both sides. It came to be known as "Stalingrad on the Yangtze".

The Chinese fought with great heroism but faced an enemy with superior weapons in all sectors, including artillery and air power. On August 23, the Japanese air force began the bombing of Nanjing, the capital, and other cities in central China. The Imperial Japanese Army had 2,700 planes, China had just 300; it had a naval fleet of 1.9 million tons, China's navy was a fraction of this size. The fighting was intense, with heavy casualties on both sides, especially the Chinese.

With the Japanese attack, like millions of other Chinese, Zhou faced agonising questions. Should he stay in Shanghai or escape and, if so, where should he go? What about his mother, wife and two young children, who lived in Suzhou? If he left, he would have to give up two jobs with good pay and conditions; would he find another one in the place where he escaped? How would he support his family?

"When the Japanese attacked Shanghai, the situation was very tense. The war developed faster than we expected. Many people told me that the foreign concessions in Shanghai had a special status and I could continue my job and earn my salary there. 'If you leave Shanghai, you will lose your job and conditions for life. So many people are escaping to Sichuan, how will you find a job there? You risk unemployment there.'"

The Japanese only attacked Chinese-controlled areas of the city but not the two Foreign Concessions. Japan had declared war on China but not on the Western powers – that would come four years later. That meant that Zhou could continue to live and work in one of the Concessions.

"After the outbreak of the war, we [Zhou and his wife] had to make a very big decision – should we stay in Shanghai or go to Chongqing [the wartime capital]. We quickly decided that we could not stay. The Japanese were very evil, especially to those who had studied in Japan. If they saw you in Shanghai, they would visit you and publish in the newspaper the next day that the Japanese commander had visited so-and-so. In this way, you would be seen as a collaborator. We decided that we could not do this and very quickly decided to go to Chongqing."

The government had chosen Chongqing, a city on the Yangtze River in Sichuan province in the far southwest as its wartime capital; without a railway line, it was too remote for attack by the Japanese army or navy – but not, as Zhou later discovered, its air force.

Zhou expected Japan to conquer Shanghai quickly, enabling it to control effectively the two Concessions. "If we stayed in Shanghai, we would actually become prisoners of war! I could only do one of two things there. One was to take part in the war; Shanghai was the front line, but I was not a soldier. The other was to continue work in the bank and teaching but I saw that this could not go on normally. I considered that Nanjing could not hold out long. So, I decided to go to Chongqing and take part in the anti-Japanese struggle there. Everyone wanted to fight the Japanese."

No Place on the Boats

Their decision to escape involved not only his wife and children, their two nannies and his mother; it also involved other members of the family – an elder sister and four children in Shanghai (the sister's husband was working in Manchuria); and an aunt who lived in the Suzhou house with her daughter. All these people became his responsibility. Then there was the question of how to reach Chongqing from Shanghai.

In normal times, they could take a boat from Shanghai; but the city had become a war zone, so it was out of the question. There were no major roads or railways to Chongqing from eastern China; and air travel was reserved for senior military and government officials. So, boat travel along the Yangtze River was the only option. Zhou planned to take the family to the port city of Wuhu in the southeast of Anhui province in central China, from where they could board a boat for Chongqing. There was a branch of the Jiangsu Bank in Wuhu; Zhou could call on its manager to help them. The decision meant abandoning the home in Suzhou where the family lived. "I was not certain I could find a job in Chongqing," said Zhou. "But I knew that there would be many there."

Departing from the Suzhou home was very painful. The family had to leave almost everything behind; they had neither the time nor the means to take items with them, only clothes and bedding. They entrusted the house and its belongings to their cook; they gave him two years of living expenses. He had chosen not to leave; like many Chinese, he preferred to stay where he lived rather than risk a dangerous journey to a destination where he had no job and knew no-one.

Wuhu was 280 kilometres west of Suzhou. On the southeast bank of the Yangtze River, it was an important river port and commercial centre for rice, tea and wood. When Zhou and his family reached there, they found the city flooded with refugees like themselves. Passenger ships going up the Yangtze toward Chongqing did not dare to stop at the main pier because of the hundreds of people fighting to board. So, Zhou had to use connections – not the sales office of shipping companies – to buy tickets for his family. They were not on a passenger ship but a vessel carrying coal that had limited space for people; it was only going to Hankou, a city up the river, not all the way to Chongqing.

Cargo vessels were all adding passengers to their load. Zhou arranged to board the ship not at the city's main pier besieged by people, but a small one on the other side of the Yangtze, in the middle of the night. Since space on the coal boat was limited, he had to buy tickets for his sister and four children, and aunt and daughter, on another boat.

They performed one other duty before they took the ship. Zhou and his wife made the 120 kilometre journey to the Anhui provincial capital of Hefei, the ancestral home of her family and her famous great-grandfather, the high official of the Qing government. His wife had been born in Hefei and lived there for the first year of her life. They visited the ancestral hall of the family to pay their respects.

Back in Wuhu, they embarked on the next stage of their odyssey. In the middle of the night, they boarded the coal carrier; it pushed off from the shore into the middle of the Yangtze. "How fortunate I was," Zhou recalled. "After all the family had settled down in the boat, my heart calmed down. The tension fell for a short while. The river was very narrow and the hills on either side very beautiful. My mood was changing rapidly and feelings were very strong. Sitting on a little boat on the Yangtze river, I felt very small."

Zhou proved extremely accurate in his forecast of swift Japanese conquests in the war. By the end of November, they had conquered Shanghai. On December 10, they occupied Wuhu, followed by Nanjing shortly afterwards. This made Zhou's decision to escape at once with his family a very wise one. If they had stayed in Shanghai and Suzhou, what would have happened to them?

The coal ship delivered the family to Hankou, the single biggest port in the middle reaches of the Yangtze. Here Zhou found a situation similar to that in Wuhu. Only one company there offered shipping services to Chongqing; it had sold all the tickets for sailings that month.

Each day a huge crowd gathered outside the ticket office; but it was closed because it had no tickets to sell. On the third day, as Zhou was queuing, a man behind him tapped on the shoulder and said: "Do you not recognise me?" He turned out to be a classmate of Zhou from St. John's University – and was deputy general manager of the shipping firm!

He was able to procure for Zhou a small cabin on a ship with space for six people – two sleeping on the bed and four on the floor. And he sold the tickets only at the listed price; on the black market outside, tickets were being sold for several times their face value. Zhou felt extremely lucky. "Our escape at that time was a fortunate meeting arranged by heaven."

The sailing was not until a week later. Zhou used the time to meet friends who had also escaped from Shanghai, and to exchange news of the war. One was a manager of the Bank of China formerly in Shanghai who had been posted to the branch in Changsha, capital of Hunan province in southern China. He said that he could not offer Zhou a job in the branch because it was too small; but a newspaper there needed someone to write daily reports and commentaries on the war. Would Zhou like the job and to make his contribution to the war effort? He believed that, while Changsha would eventually fall to the Japanese, this was not imminent. He, and other friends, advised Zhou to settle his family in Chongqing first and then return to a city closer to the front line where he could do something useful. He accepted their proposal; he would take his family to Chongqing and then go alone to Changsha.

When the family arrived in Chongqing, they were very fortunate, compared to the thousands of other refugees reaching there. Zhou's wife had a classmate who had a relative living in Hechuan (a small, quiet town north of Chongqing). The relative welcomed Zhou's mother, wife, children and two nannies into his home; he later helped them rent their own house in Hechuan."It was a small place. The Japanese planes were not likely to attack it. They would certainly bomb Chongqing. When I arrived there, this bombing had not started. But they would do so constantly. Hechuan was also close to Beibei, a place with lovely scenery and hot springs. It was a very suitable place to settle the family. Once they were there, I felt at ease. If they were in Chongqing, I would not be."

A Journalist in Changsha

After he settled his family, Zhou went alone to Changsha, capital of Hunan in southern China. He found it packed with refugees from Shanghai, Nanjing and Hankou. He took the job recommended to him by his friend, the manager of the city branch of the Bank of China. It was in a small newspaper called *Li Bao* (the Journal of Strength). The editorial office consisted of a large room with two beds; he slept in one and the other editor in the second one.

His job was to write articles and commentaries, 1,000-2,000 characters long, on the war with Japan, including how people could contribute to the fight. He received no salary, just board and living costs. With a large refugee population, Changsha was a centre of news and information, a good place for a journalist to work. While Zhou had not worked full-time as a reporter before, he had plenty of experience in writing, including university essays, articles for student publications and reports for Jiangsu Bank. He did not use his real name, Zhou Yaoping, but Zhou Youguang, the name he later adopted and by which the world knows him today. The articles were widely read by those living in the city; they were hungry for information about the war, about whether they could go home and when or if the Japanese army would attack Changsha. He was making a contribution to the war effort.

Since sleeping in the newspaper office was not comfortable, Zhou's friend at the Bank of China found him a room in the city YMCA. This had two great advantages. The city suffered from limited power and water supply – but the YMCA had built a well and a wind power generator on the roof, so that the rooms were well lit and had running water. The city's telegram system worked well, so he was able to stay in touch with his family in Chongqing. After he had been in Changsha for a month, Zhou met a friend from Hankou. His advice was that, although he was doing good work there, Zhou could make a greater contribution in Chongqing; as the wartime capital, it was becoming the centre for national resistance. Go there and he could do even more.

Another factor to consider was the ability to reach Chongqing before the Japanese made this impossible; as he had discovered with his family, this was no simple matter. After the Japanese military captured Nanjing in December 1937, its next target was Wuhan, the second largest city in China, halfway upstream the Yangtze River, with a population of 1.5 million. It was the major river and railway hub of central China and, for Zhou, the best place to catch a boat to Chongqing. He knew that, in the first half of 1938, the Japanese military was preparing its attack on Wuhan, and that it was likely to succeed. All this persuaded him to decide quickly and make the journey to Chongqing, while Wuhan was still under Chinese control.

After a month in Changsha, he returned to Hankou, part of Wuhan. There he found even more refugees than during his first visit. He could not find his friend in the shipping company and relied on a friend in a bank to procure a boat ticket for him. The voyage to Chongqing was not simple, with stops along the way and one boat with limited horsepower so that it could only travel during the day and not at night. As before, all the boats were crowded with people escaping. It was a great relief when he finally reached Chongqing again. These were his thoughts after his arrival: "With such a backward place as Sichuan, how could we defeat Japan? That would be extremely difficult and we needed other ways. Although China did not have the strength to defeat Japan, I was certain that we would win. The war would expand; the situation in Europe was very tense. Everyone said that Japan would attack China and China would not surrender. It would retreat and continue to resist. More and more countries would join us against Japan and Germany." His forecast proved to be extremely accurate.

On June 15, 1938, the Japanese army launched its attack on Wuhan. The battle lasted four months, ending with a Japanese victory on October 27. Both sides suffered enormous casualties; the Japanese victory was only possible due to use of chemical weapons, which was condemned by the League of Nations.

The heroic resistance of the Chinese soldiers greatly weakened the Japanese army and enabled equipment and thousands of troops to be moved to the southwest, to prepare for a long war of resistance.

Chapter Four

World War – Feeding The Nation, Escaping Bombs

On his return to Chongqing, Zhou found the city was being transformed after the government had officially moved its capital there in November 1937. In 1932, it had a population of 269,000. By the end of the war, this had more than tripled to over a million, with the arrival of thousands of civil servants, military officers and workers in more than 1,000 factories which had moved from the east.

Chongqing was poor and backward, with no rail or proper road links to the rest of China; people and goods all had to use the Yangtze River. It was this very remoteness that made it attractive as a wartime capital, beyond the reach of the Japanese army. To someone like Zhou, used to the pace of life in Shanghai, things moved slowly in Sichuan. Workers took frequent breaks for a smoke, a chat and a snack; on a normal day, they did no more than six hours of work.

Industrialists from the east transferred their factories and rebuilt them on the outskirts of Chongqing and in neighbouring towns, just as the Soviet Union moved factories east of the Ural mountains after the Nazi invasion of June 1941.

Among the first people Zhou encountered in Chongqing was a man who had received the first accounting licence in China, from the Beiyang government (1912-28). He had taught accounting at Kwang Hua University in Shanghai, where Zhou met him. He invited Zhou to go to Chengdu, capital of Sichuan province and 300 kilometres from Chongqing; he asked Zhou to help set up a new campus of Kwang Hua University, as well as a secondary school.

A well-known figure in China, the accountant was able to raise money for the project from wealthy people in Sichuan. Zhou agreed, and his wife went with him, as did other colleagues from Kwang Hua in Shanghai.

Both started teaching there and considered settling in the city. Living conditions were better than in Chongqing; it was less likely to be a target of Japanese bombing.

But other friends urged Zhou to return to Chongqing and join the Agriculture Bureau (AB), a unit of the Ministry of Economics. Set up in 1936, it was responsible for production, storage and sale of grain, cotton and textile fibres. It was headed by a Vice Economics Minister. Zhou's friends told him that, if he wanted to play an important role in the anti-Japanese war, this was his chance.

The AB was a department of the central government responsible for commodities critical for the war effort. As the wartime capital, Chongqing was the centre of national resistance. Zhou was persuaded and moved to Chongqing. His wife chose to stay longer in Chengdu, because it was less likely to be bombed by the Japanese.

The AB had two main objectives – feed the people of southwest China and provide them with cotton, clothes and textile fabrics. It included a financial arm, to provide farmers with the money to buy seeds, fertiliser and other production materials, as well as technical assistance. Its area of responsibility covered the four southwest provinces of Sichuan, Guangxi, Yunnan and Guizhou, with a combined population of 88 million; they were too remote for the Japanese army to reach. Zhou found that the senior staff of the AB were high officials who had moved from Nanjing, Shanghai and other cities under occupation. It was not like post-1949 China, whose government took over and controlled the land. The farmers continued to own the land; the AB only provided funds, support and technical help.

Zhou was appointed deputy commissioner of Sichuan province, in charge of dozens of co-operative banks; these were set up in more than 30 of Sichuan's counties. Most of the managers were graduates of the Department of Agricultural Economics of Jinling University in Nanjing, who had, like Zhou, moved to Chongqing. The new job was a heavy responsibility for someone just 32 years old and with only three years of banking experience.

In 1938, Sichuan had 53 million people, the largest population of any province in China. Zhou's job was not only to ensure the livelihood of millions of farming families and the output of one of China's most important agricultural areas. Since Chongqing was the headquarters of the government, Sichuan's meat and grain fed China's soldiers, sailors and airmen; their clothes and blankets were made from its cotton and textiles.

This is how Zhou summarised his contribution there. "I worked there for several years. This can be considered my work most directly related to the anti-Japanese struggle. During a war, the most fearful [thing] is [to] have no food to eat nor clothes to wear. We borrowed the American method, through finance, to help landowners and farmers maintain production of grain and cotton. As a result, during the eight years of war, from what I saw of this work behind the front, there were no shortages of grain or cotton. We can say that this was no simple matter and that it was a success." He Lian, his boss, thought very well of him: "Zhou was a very capable person. A university graduate, he had worked as a deputy general manager at a bank in Shanghai. He had rich experience."

Sichuan was, and is, one of the most important agricultural provinces in China. But, while its land was good and farmers were skilful, transport, storage and distribution were poor. Many crops were grown on mountains and transported over dirt roads without vehicles; they were carried by horses, camels, mules and people. The AB helped with transport, processing, packaging and storage; it built warehouses. The area had few railways, so the AB carried goods in trucks and boats. Its technical staff helped farmers fight pests and insects.

Zhou went from one county to another to inspect the work of his colleagues and farmers. "In small places, I saw the gap between rich and poor and the good life enjoyed by landlords and feudal warlords. Many ordinary people had no job and nothing to eat. In the beginning, I often saw on the side of the road corpses of those who had died of hunger. Thanks to the enormous economic work of the National Government in Sichuan, by the second year, the economy had improved and I saw no such corpses. The main task of the AB was to guarantee production of grain and cotton – food and clothing. During the anti-Japanese war, no major problems occurred in these two in the rear areas."

Below the Bombs

As Zhou had predicted, his work and that of his colleagues was soon interrupted by Japanese bombing of Chongqing and surrounding areas. The first such attacks on civilian areas in history had occurred during the Spanish Civil War of 1936-39, by the air force of the Fascist side. This practice was followed by the Japanese, German, American and British air forces in World War Two and, most recently, by the Russian air force during its invasion of Ukraine from February 2022.

The aim was to demoralise the civilian population and put pressure on the government to surrender. Between February 1938 and the end of 1944, Japanese bombs killed 32,829 soldiers and civilians in Chongqing. They dropped more than 11,500 bombs and destroyed over 17,600 buildings. The Chinese air force was ill-equipped to defend the city against the most advanced fighter planes in the world, Mitsubishi G3M bombers and Zeros. Between 1938 and 1941, the city's air defences shot down about 100 Japanese planes. The residents built a vast network of tunnels where they took shelter during the raids. The bombing failed to achieve its objective. It did not destroy the morale of the residents – it only made them even more determined to resist.

Zhou remembered three raids in particular. One day, he left the city with his AB colleagues to visit a village and see the farming there. When they returned, they found that their office had been destroyed; nothing was left. On another occasion, the Japanese used firebombs; these were very effective because most houses in Chongqing were built of wood. Fires swept through the city. The next day people climbed to the top of the hills on which the city was built and saw slopes completely bare, with everything burnt to ashes. One evening, on his way home from work, he was walking down a slope when a Japanese plane came toward him. "The explosion threw me into a ditch. I did not know if I was alive or dead. I did not dare to move. I waited for the plane to fly away before I got up. I thought I would be injured, but I found no pain anywhere on my body. The people around me were all dead."

To avoid the bombs, Zhou moved his office and home several times to remoter areas outside the city. During their eight years in Sichuan, the family moved a total of 36 times. The first time, they moved the office to Yibin, a town upstream the Yangtze. Since the boat there from Chongqing was slow, he often boarded a small plane for two passengers to the city for meetings. It flew close to the river, so close that he could see the fish; his wife called it "the dragonfly plane". The plane helped to save the life of his wife when she suffered a severe case of dysentery. Since Yibin had no doctor, the "dragonfly" brought a well-known doctor from Chongqing to treat her. Thanks to his excellent care, Madame Zhou made a full recovery.

Other officials followed Zhou to Yibin; soon the Japanese learnt of this. So they bombed the town too. The same thing happened to other places where Zhou moved; the Japanese reacted most swiftly when they learnt that senior members of the government had moved to a place. They were prime targets of the bombing. The Zhous were happiest in a town called Jiang'an, where they lived in the house of an intellectual, a house with many rooms. Zhou's son, then five years old, was able to attend the local school. The town was home to a Drama School; one of its teachers was Cao Yu, one of China's most famous playwrights. Zhou and his wife became close friends with Cao and his family. After 1949, the school became the Central Drama School in Beijing.

During the war in Chongqing, as well as in Kunming and Chengdu, lovers of Kunqu came together for performances; it was a way to forget, for a short time, the terrible war going on around them and to meet fellow lovers of the art. One tragedy that occurred in Jiang'an was the sudden death of the older of the family's two nannies. She contracted malaria. The family took her to the local missionary hospital; but it was unable to cure her and she died three to five days later. "It was a very great shock to us. We had a deep feeling for her," Zhou wrote.

As the war went on, it became more costly. The government had to print more money to finance it – causing inflation. This badly affected civil servants like Zhou, who had to dip into their savings to make ends meet. It also diminished the value of the AB, whose loans to farmers were worth less. In relative terms, farmers were better off – their assets were not money but grain, cotton and other commodities whose value did not fall.

During his time in Chongqing, Zhou met Zhou Enlai, one of the top Communist leaders, as well as senior members of the ruling Kuomintang. Both parties invited him to join, but he declined. He preferred to keep his independence as a scholar. After 1949, he would meet Zhou Enlai again, in Beijing, as well as other Communist leaders, including Chairman Mao and Chen Yi, who served as Mayor of Shanghai and later Foreign Minister.

Back to the Bank

One day, in Chongqing, Zhou met a man who introduced himself as Xu Boming, general manager of the Bank of Jiangsu, his former employer; he had also escaped from Shanghai. The Japanese had occupied the head office in Shanghai, like those of other banks there. Xu had set up an office of the bank in Chongqing. It had little work to do; its mission was to prepare for the return to Shanghai after the victory. Xu asked Zhou to run the office; he replied that he was employed at AB. "No problem," said Xu. "You can do both jobs".

His condition was that Zhou lived in the apartment the bank had rented as its office. With AB's agreement, Zhou moved into the apartment and took up the post. The AB work occupied most of his time. Xu was a great lover of Kunqu and Beijing Opera and supported its performers. Zhou soon had a new roommate in his apartment – a prominent young Beijing opera singer, who was also a distant relative of his wife.

On Fridays, Zhou attended a regular meal with other members of the Shanghai financial and industrial elite who had moved to Chongqing. He was impressed by the people he met – capitalists who had, with the encouragement of the government, moved their operations to this remote and ill-equipped city in southwest China. It had few of the facilities, transport, skilled workforce and raw materials to which they were used in Shanghai. "In my view, they came to Sichuan, built many factories and produced many items. They made a contribution to the anti-Japanese struggle," Zhou said.

Mother Moves to Rangoon

As the bombing intensified, it became harder and harder to find a secure place to live. Zhou was especially concerned about his mother. She had lost her home in Suzhou and made the hazardous trek to this remote corner of China; but she could not find a safe place to lay her head at night. He thought of his elder sister, Zhou Run. She had taught in Singapore and became principal of the Overseas Chinese Ladies Secondary School in Rangoon, capital of the British colony of Burma. From Shanghai, Rangoon was remote and inaccessible. But, from Chongqing, it was suddenly close – there were regular flights there. The Allies supplied the Chinese government with arms and materials from bases there via the Burma Road, completed in 1938; it linked Lashio, at the end of a railway from Rangoon, to Yunnan.

At the time, Burma was at peace. In 1941, Zhou decided his mother would be safer there than in Chongqing; so he put her on a plane to Rangoon, to stay with her daughter. After she had settled, Zhou's sister invited him for a visit. After receiving permission for a holiday from AB, he took one of the many planes to Rangoon. He found it a peaceful and prosperous city, without Chongqing's constant stress and air raids.

He found a society with three "castes" – at the top were the British, the colonial power. Their firms dominated export of Burmese rice and other high-quality products, from which they earned a large profit. The second "caste" were Chinese and Indians, who dominated wholesale and retail businesses. In addition, the Chinese owned 90 per cent of the factories that milled the rice, using modern British equipment. At the bottom were the Burmese farmers who grew the rice. The Chinese earned a modest profit and the Burmese very little.

The large Chinese population was divided by language; there were three kinds of schools – in Cantonese, Fujian and Hakka – set up by their respective communities; they had little interaction with one other. The May 4 Movement in China had provoked a wave of nationalism and realisation of the importance of a common language, Mandarin. Zhou's sister was a beneficiary of this; she wrote plays in Mandarin and promoted them among her students.

Zhou found that the Chinese consul in Rangoon was a classmate from St. John's. Through him and his sister, he had the connections to arrange a comfortable job in Rangoon. His sister said: why not settle here and bring the family? But Zhou understood geopolitics too well. He realised that, given Burma's importance as a supplier of arms and material to China, a Japanese invasion to cut off this route was only a matter of time. So he declined the offer. He was proved right within less than a year.

In addition, Zhou learnt something valuable in Rangoon – the value of Pinyin, which means spelling in Roman letters. He found that, while the Burmese were poor, few were illiterate. This was because, as children, they first learnt their language in Pinyin, usually in Buddhist temples. Then they learnt the Brahmic script in which Burmese is written; it is also used for the liturgical languages of Pali and Sanskrit. Within six months, they could read simple texts. It was an important lesson Zhou took back with him to China.

After his return from Rangoon in 1941, Zhou decided to leave the AB and take a job with Xinhua Bank. This followed a meeting with Wang Zhishen, general manager of Xinhua Bank. Its head office in Shanghai had been a short distance, on Jiangxi Road, from that of Bank of Jiangsu.

The Xinhua Bank branch in Chongqing was very active. Why did he leave AB? One reason was the high inflation that badly affected civil servants on fixed salaries; the bank offered an attractive salary, and a car. Another was conflicts within the government that persuaded He Lian, Director-General of AB and Zhou's boss, to resign. A third was that working for the AB involved constant travel around Sichuan, which was difficult and exhausting.

By contrast, work at the bank was simpler, less political and what Zhou had been used to before the war. One good outcome was that Zhou was able to move his family to Chengdu, where Xinhua Bank had a small office. It was safer and suffered less bombing than Chongqing; his wife resumed teaching there, at a secondary school under Kwang Hua University. For his work, Zhou had to continue living in Chongqing.

World War

On December 7, 1941, the Japanese launched their attack on Pearl Harbor, triggering a war with the United States and a global conflict. This made an attack on Burma merely a matter of time, as Zhou had predicted. Zhou's sister sent their mother back to Chongqing. His sister remained there on her own. Chinese in Burma escaped to the border with China, and from there to Kunming, capital of Yunnan province.

On December 22, the Japanese army crossed the border into Burma and made steady progress. The British army there was ill-equipped and under-manned. On March 7, 1942, it abandoned Rangoon after setting the city on fire; the Japanese took over. Their victory closed the Burma Road. From then on, the Allies had to supply China from India by flying arms and materials over the Himalayas, a long and hazardous route. The supply of material diminished.

Back in Chongqing, Zhou followed these developments with intense interest. He had an encyclopaedic curiosity which remained with him throughout his life. "Britain then was still weak. We followed the repeated victories of Germany, one after another. It greatly alarmed us in Chongqing, especially the fall of France. These failures in Europe surprised us even more than the advances of Japan in China and made us feel that this war would last for a long time. But everyone was confident of China's victory – that was the strength of the nation."

He said that Japan made two enormous mistakes. One was to wage an unlimited war against China and the other to attack the United States; the U.S. was a country with a substantially larger industrial capacity than Japan to produce warplanes, naval vessels, tanks, artillery and other armaments. "After the Pacific War broke out, we in Chongqing felt elated, moving from depression and perplexity to a feeling of certainty. We put our hopes onto the Americans. Initially, they concentrated on the war in Europe ... the U.S. decided to finish the war in Europe before finishing the war in Asia. This meant that China would face great difficulties."

Zhou was not alone in believing that Japan had made a fatal mistake. Isoroku Yamamoto, the admiral who led the successful attack on Pearl Harbor, agreed. He had opposed his country's occupation of Manchuria and all-out war on China. After two postings as a naval attaché at the Japanese embassy in Washington and speaking fluent English, he knew better than anyone in his government the industrial capacity of the country they had chosen to attack.

In mid-1941, then Prime Minister Fumimaro Konoe asked him what would happen in a war with the U.S. He said: "I shall run wild considerably for the first six months or a year, but I have utterly no confidence for the second and third years." This is exactly what happened. For the first six months of the Pacific War, Japan conquered territories and islands in Asia and the Pacific. Then, in June 1942, it suffered a major defeat at the Battle of Midway; this tilted the balance of power in the Pacific towards the U.S. and it never lost the initiative from then on.

After the attack on Pearl Harbor, the Japanese military invaded Guizhou, one of the four provinces of the southwest under the control of the Nationalist government. There was panic in Chongqing and talk of moving the capital again, to Xichang, capital of Xikang province, 700 kilometres to the south. (Xikang was a province created by the Republic of China in 1939, one that existed until the arrival of the People's Liberation Army in 1950). But the Japanese retreated from Guizhou; the capital remained in Chongqing.

In June 1941, in Zhou's view, Adolf Hitler had made a mistake as fatal as that of Japan in attacking the United States – he invaded the Soviet Union. The scenario followed that of Japan's attack on China – rapid victories and the retreat of the government and much of its industry into the interior. Like Japan, the Nazis won battles but could not win the war. The terrible winters, the rebuilding of Russia's industrial power and the determination of its people turned the tables.

The Chinese looked on with increasing confidence as the war turned in their favour. The Americans began the bombing of the Japanese mainland; but Japanese bombers could not reach Houston or San Francisco. In November 1943, President Chiang Kaishek went to Cairo to attend a conference with U.S. President Franklin Roosevelt and British Prime Minister Winston Churchill. They discussed the post-war settlement in Asia. It was the first time in history that a Chinese leader had been invited to attend the "high table" of great powers in the world. "This meeting made the people of China believe that victory was 100 per cent certain," said Zhou. "The joy and happiness of everyone at that time was indescribable."

Loss of Beloved Daughter

Millions of Chinese perished during the anti-Japanese war. One of them was Zhou's young daughter, Zhou Xiaohe. In 1942, aged six, she died in Chongqing of peritonitis because she did not receive the medicine available to her in Suzhou and Shanghai. "Had there been no war, she would not have died," said Zhou. "During the war, medical conditions were extremely difficult. To lose a little daughter, there are no words to express the pain." The loss also devastated her mother. For her, the eight years in Chongqing were a Calvary – constant moving from one home to another; fear of Japanese bombing; caring for her children, mother-in-law and other relatives; the regular absence of her husband away for work and being unable to contact him; and the absence of proper medical care in the rural areas where they often lived. Like her husband, she counted herself very fortunate to survive the war unscathed, while so many around them died or were wounded.

Shanghai was the principal manufacturer of medicines in China. With the Japanese occupying the eastern part of the country and the National Government the southwest, it became very difficult for doctors in Chongqing to procure the medicines they needed. It involved a tortuous road journey from Shanghai to Jinhua in Zhejiang province, which was under the control of the National Government; from there, the medicines went to Chongqing. The journey took too long for doctors themselves to make the trip; so merchants did it for them, buying them cheaply and selling them in Chongqing for a healthy profit.

By 1942, inflation had reduced the value of Zhou's salary. So, in April that year, he accepted the offer of a doctor to accompany him on a trip to Jinhua, to buy medicines and resell them in Chongqing. While he was there, he met James Doolittle, one of the most famous American pilots of World War Two. On the night of April 18 that year, Doolittle led 16 bombers over Tokyo; it was the first American bombing raid over the Japanese capital. While it caused minimal damage to military and industrial targets, it had a major psychological effect. It demonstrated to the Japanese public that their air force could not protect them; it also greatly lifted the morale of Chinese people. After the raid, the pilots flew westward and landed in China, where they donated their aircraft to the government.

A military officer in Jinhua who knew Zhou invited him to his house for dinner, to serve as interpreter between Doolittle and his fellow pilots and the Chinese officers hosting him. Zhou joked with them that Doolittle's name was misleading; he had done a great deal. Over the next few days, Zhou rode with Doolittle and the other pilots in a jeep to Guilin, capital of Guangxi province in the south. From there, Doolittle was flown to Chongqing; Zhou had to take a bus. Doolittle went on to hold senior posts in the U.S. Air Force in the Mediterranean, North Africa and Europe for the rest of the war. In 1946, when he was working in New York, Zhou visited Doolittle in his sumptuous office at the Shell Oil Company, where he was a vice-president. It was after Zhou returned to Chongqing that he learnt the tragic news of his daughter's passing.

New Branch in Northwest China?

In 1943, Zhou was one of five bank representatives sent to northwest China to look at business opportunities. It was an area of the country little touched by the war. Since it was poor and undeveloped, Japan had no interest in occupying it. It was the base for the Communist Party and its military forces during the war.

At the request of his superior, Zhou lived for a period with a family in Xi'an, the main city in the area; as Chang'an, it had been capital of China during several dynasties, including the Tang.

Zhou was not impressed by what he saw. He found the region poor and backward, with a harsh climate, including low rainfall and fierce winds; its residents were inward-looking and protectionist. Many conducted business through the exchange of goods, not money. So, his superior decided against opening a branch there and summoned him back to Chongqing.

The people who impressed Zhou most were French and Italian Roman Catholic missionaries stationed in the northwest. They had learnt Mandarin and the local dialects and knew more about the economy and conditions than many of the local officials. Zhou admired them for providing education, health care and medicine to the local people. "They went to remote areas, to the villages and mixed with ordinary people. They understood very well the situation in China and sent regular reports to Rome. Today, to develop the northwest, we need the organisation and support like that of the church and we need people willing to endure hardship and sacrifice themselves in their work."

Secret Visit to Shanghai

As the Americans continued their advance across the Pacific and the war turned against Japan, the bankers in exile in Chongqing began to plan their return to Shanghai.

In Chongqing, Zhou received two new assignments from his boss. One was to set up and manage a trading firm to bring to Sichuan goods made in Shanghai, especially clothes and socks. This was most profitable. The route would be from Shanghai to Jishou in Hunan province and from there to Chengdu, for distribution all over the southwest. Zhou agreed to this. The second assignment was more difficult – go secretly through Japanese lines and visit Shanghai, to meet the manager running Xinhua Bank's operations there. Zhou accepted this task also.

He was able to reach Shanghai and meet the manager. He found things not as he expected. The Japanese had not closed the banks there but allowed them to continue operating. After the start of the Pacific War, no imports arrived in Shanghai. This was a good opportunity for Shanghai manufacturers to increase their sales. They and the trading firms were the main customers of the banks. "In this respect, the Japanese were very intelligent," said Zhou. "They did not shut down many financial institutions. Since they were operating within the (Japanese) cage, what did they have to fear?" The manager gave Zhou an update, saying that the banks were running much as before 1937 and that he was waiting for the head office to return to Shanghai after the war. Zhou did not dare tell his friends or relatives in the city about his visit, for fear the Japanese would find out.

The head office of the new trading firm was in Chengdu, so Zhou moved there to manage it, with his family. They rented a pleasant one-storey home with a garden. The living environment was more pleasant than that of Chongqing, with less risk of bombing. Chengdu was the historical capital of Sichuan, with a rich culture and history. But even there they could not escape the war. One day his young son was playing in the garden, when a stray bullet hit him in the stomach. Fortunately, the city was home to a military hospital serving a U.S. air base in the city; doctors there removed the bullet. His son kept it for years after as a souvenir of the war.

Their stay in Chengdu did not last long. On August 6 and 9, 1945, the U.S. air force dropped atomic bombs on Hiroshima and Nagasaki; Emperor Hirohito surrendered unconditionally on August 15. "At the time of the surrender, I was in Chengdu. Our excitement was indescribable. We quickly moved back to Chongqing and prepared to return to Shanghai," Zhou said.

CHAPTER FIVE

United States And Britain – Learning Every Day

With the end of the war, the tens of thousands of exiles from Shanghai, Beijing, Nanjing and elsewhere in Chongqing were impatient to go home. Many did not know what they would find nor if their homes were still standing and their belongings inside them.

The obstacles of travel were greater than in 1937, because everyone wanted to leave at the same time. The only options were a boat on the Yangtze or, for the privileged elite, an aeroplane. Zhou had to wait until the end of 1945; he was fortunate. His boss, Wang Zhishen, was eager to get the headquarters of Xinhua Bank in Shanghai up and running as soon as possible.

Wang used his connections to secure Zhou a place on an American military plane flying troops from Chongqing to Shanghai. Since it was not carrying weapons, it had empty space – but no seats and no meals. So Zhou sat on the floor: it was uncomfortable, but the fastest and most direct route. He had a meal before the flight to ensure he travelled on a full stomach. He went on his own, without his family.

When he arrived in Shanghai, he felt that he had arrived on another planet. After reaching the bank, the first thing he did was to take a shower and have a glass of clean, running water. The water of Chongqing was muddy. There was no clean running water – vendors earned a brisk trade selling bottled water.

His family – mother, wife and son – had to take the same route as everyone else, by boat down the Yangtze to Wuhan. They did not board a passenger steamer but a wooden boat. It was extremely dangerous, especially passing through the Three Gorges; many boats capsized. The three travelled with relatives and friends. En route, one of the family members fell into the Yangtze and drowned. The others arrived safely in Wuhan. From there, with the aid of Zhou's friends in the financial community, his family had a comfortable journey to Shanghai.

When they visited the family home in Suzhou, they found that everything in the house had gone; nothing was left. "When we had left, we had expected to be away for no more than three years. In fact, it was eight."

General manager Wang and his colleagues had also arrived in Shanghai from Chongqing. They wanted to resume and expand business. Wang decided to send a senior staff member every two years to the United States to work, study and learn modern methods and equipment and how to use them in China. High on the list was accounting machines, to replace accounts written by hand by staff in Chinese banks; the machines could calculate the accounts and print them out.

 "After 1949, these machines were rejected and not used again. It was a great pity," Zhou wrote. He was an important part of Wang's plans. In New York City, Xinhua Bank had had an office for many years on Broadway. In 1946, Wang sent Zhou to run the office there, and open a branch of the trading company he had managed from Chengdu.

Following their close co-operation during the war, China and the U.S. had very good relations. It was easy for Zhou to obtain a visa for himself and his wife. He sent to the consulate in Shanghai a family photograph and proof that he had US$4,000 in a bank account in the U.S. They decided not to take their young son; they sent him to the home of Zhou's brother-in-law in Suzhou, where he was principal of the secondary school established by his father. They wanted him to continue his education in a Chinese school and a Chinese environment.

Since only military planes were flying across the Pacific, Zhou and his wife

Zhou Youguang and Zhang Yunhe set sail for the United States from China Merchants Wharf in Shanghai.

in 1946 took a mid-size passenger liner from Shanghai to San Francisco; she was seasick nearly all the way and stayed in the cabin. When she felt better, she sang Kunqu songs with a friend who was also a fan of the genre; it greatly lifted their spirits.

On January 13, Zhou marked his 41st birthday; that day the ship crossed the International Dateline, so he was able to celebrate his birthday twice. The liner made no stops and arrived in San Francisco after only two weeks. Zhou and his wife stepped on American soil for the first time. They took a train across the continent to New York. Zhou was both the representative of Xinhua Bank there and the head of the trading company. He registered the firm, to make it an American company, and hired two people to serve as deputy general managers.

The banking business was mainly conducted through their business partner, Irving Trust Company. It was founded in 1851 as the Irving Bank of the City of New York; it changed its name to the Irving Trust Company in 1929, when it was the fourth-ranked financial institution in New York and the fifth-ranked in the U.S. Its headquarters was at One Wall Street.

Zhou's first duty was to manage the bank's office in New York. His second was to learn American banking practices and hardware and decide what he could bring back for use in China. He attended economics classes at Columbia and New York Universities; he made it his routine, after dinner, to go to the New York Public Library and read from its treasure house of books, especially on economics. He usually stayed there till closing time at 22:00. "It was the happiest thing I did during my time in New York," he recalled.

He also explored the city's many museums, including the one on Natural History. This study was another example of his encyclopaedic curiosity and eagerness to learn many things, including those unrelated to his work. "I went to the U.S. to work, not to study. My conditions of work were very good. I received money from both the Chinese and the American bank. I was a high-level officer and received expenses for travelling. After we returned, I wrote a report on the travels. In my spare time, I studied and was very diligent. I did not want to waste any time. The conditions for study in the U.S. were excellent. I researched economics." He so impressed the staff of the Public Library that they lent him the use of a small study room to do research.

During the summer, the couple visited the University of Michigan at Ann Arbor to attend a summer school and meet one of Zhou's sisters-in-law and her husband who were living there. One speaker at the summer school was Chao Yuenren, a famous scholar of linguistics and also a native of Changzhou. Chao presented his plan for a romanised version of Chinese, which he called Hanyu Pinyin. Zhou found the system very good; it greatly influenced his work in this field 10 years later.

Born in Tianjin in 1892, Chao was one of the most gifted linguists of his generation. In 1910, he went to the U.S. on a Boxer Indemnity Scholarship and studied at Cornell University. He earned a Ph.D. in philosophy from Harvard University in 1918 and later taught there. He spoke French and German fluently and some Japanese. After teaching posts in China, he returned to the U.S. in 1938 and lived there for the rest of his life. In 1945, he served as president of the Linguistic Society of America. We will describe his work in more detail in Chapter Six, during Zhou's development of the Pinyin system.

An important linguistic initiative was under way at Harvard-Yenching Institute at Harvard University – a dictionary of Chinese words from ancient times to the present. Zhou had a friend involved in the project, Li Fangkuei. It was very ambitious, with detailed explanation of the meaning and background of each character, and would take several years to complete.

"I was extremely interested in linguistics, although I had not done research in this area," Zhou said. Meeting these scholars stimulated his interests in linguistics, especially the study of Mandarin. Another important Chinese he met in the U.S. was Lao She, one of the most important Chinese novelists of the 20th century. He was in the U.S. on a two-year cultural grant sponsored by the State Department, lecturing and overseeing the translation of several of his novels into English. Zhou remembered him as a very interesting man, full of jokes.

"In the United States, most of my friends were Westerners, not Chinese. This was intentional. Through knowing Westerners, I could understand American society, which would be especially helpful toward my work in finance."

He met people at the top of society, including government officials and directors of large companies. They invited him to a box at the New York Metropolitan to hear Italian opera. His work enabled him to meet heads of large manufacturing companies, including makers of flour and textiles.

One of his missions was to help clients in China acquire advanced equipment and machinery for their factories. One that interested him greatly was Remington Rand. Formed in 1927, it became the largest producer in the U.S. of typewriters, a piece of machinery vital in making business faster and more efficient – and, at that time, there was no typewriter to write Chinese. After World War Two, it became one of the country's largest computer companies.

During the war, it was contracted by the government to manufacture arms; it produced a .45 calibre semi-automatic pistol for the army and made more of a particular model than any other company. For Zhou, this was one reason why the U.S. had become such a military power during the war – it was able to direct private companies to make weapons and increase production of them rapidly.

Many things about the United States impressed Zhou. One was the high efficiency of office work, with widespread use of telephones and telegrams and managers and secretaries working closely together. He also saw shorthand as one reason for this efficiency. There were two systems – Pitman from Britain and Gregg from the United States.

Zhou himself developed a system of shorthand symbols for the four most common dialects of China – Beijing, Shanghai, Guangdong and Xiamen. He even went to meet John Robert Gregg, creator of the second system, then in his 80s.

Zhou did a great deal of research and created a useable system. But he did not make this public; he feared that society would not accept it – such was people's deep attachment to characters. "I kept all my notes until after the foundation of New China (1949). But, during the Cultural Revolution, I destroyed them all," he said.

Another reason for the economic success of the U.S., he believed, was the intense competition between companies that created new products and shut down firms that did not. A third was the financial system, with stock and bond markets working efficiently to attract capital from the public and allocate it for investment. A fourth was the railway system. On holidays, he and his wife rode in luxury cars with sofas and large windows which took them all over the country. This trans-national system had been a key reason for the country's rapid economic growth and China should copy it. "The peak of railway construction was before 1914, with more than 30,000 kilometres laid in a single year, more than the entire network in China at that time. All the railway companies are privately owned," he said.

Zhou lived in the U.S. at the end of the golden age of its railways. But, between 1945 and 1984, non-commuter rail passenger traffic declined by 84 per cent. A railroad country became an automobile country. Millions moved from the cities into suburbs, where residents needed cars to reach their homes. They bought goods, ate meals and exercised at suburban shopping malls, not the city centre. The Federal government poured billions of dollars into a nationwide interstate motorway system and imposed severe taxes on the railway companies.

During his stay in the U.S., Zhou had the good fortune to meet Albert Einstein, one of the greatest physicists of the 20th century. A Jew from Germany, he settled in the United States after Adolf Hitler came to power in 1933. In 1939, he warned U.S. President Franklin Roosevelt that Nazi Germany was researching how to make atomic bombs, one reason why the U.S. began similar research.

Zhou received an introduction from a friend and his former superior at the Agricultural Bureau in Chongqing; he had become a visiting scholar at Princeton University after World War Two. Einstein was teaching there at the Institute for Advanced Study. "Before the meeting, I thought: 'What should I talk to him about? It was very sensitive to discuss the atomic bomb. So, I would not mention it unless he brought it up.'" In the end, they spent an hour discussing Sino-US relations and the situation in China. "He was easy to approach and spoke slowly and softly. He had no airs at all. It was a great honour to meet him."

During his two years in the United States, Zhou reaped a rich harvest. He used the time and opportunities offered by a high-paying job to make many American friends and learn a great deal about their rapidly developing country.

Yellow Flour as Egg

In 1948, Xinhua Bank ordered him to move from New York to London. His mission was to prepare to resume services with the Bank of England which the bank had started before World War Two, but had been forced to postpone. These were foreign exchange remittances and export and import mortgage loans.

Just as it had been going to the United States, it was easy to obtain a visa for Britain. He and his wife took the Queen Elizabeth, a luxury liner and the largest passenger vessel ever built; launched in 1938, it held that record for 56 years. It plied the route between New York and Southampton from October 1946 until the end of 1968. It carried 2,300 passengers and 1,000 crew, at a speed of 59 kilometres per hour; it was 314 metres long and 36 metres wide.

Most passengers were Americans going to and from Europe for business or pleasure. After World War Two, Britain was very short of foreign exchange. The liner, and its sister ship Queen Mary, earned large amounts of this precious foreign exchange.

Zhou marvelled at the size and opulence of the Queen Elizabeth. Passengers were divided into three classes – he and his wife bought tickets for a second-class cabin. There were seven storeys above the water level and four below, with lifts for passengers and for goods. During the evenings, Zhou put on a formal suit and ate dinner in a large hall on the first-class floor. It had a band, to which Zhou and the other passengers danced.

Each class had its own small bank, swimming pool and department store; the goods were cheap because they were not subject to American or British taxes. Unfortunately, as during the crossing of the Pacific, his wife was seasick and took her meals in their cabin; on the dance floor, Zhou had to find other partners.

He found Britain a great contrast to the United States. While it had won the war, it had paid an enormous price, in loss of manpower and destruction to its cities and industries. The value of sterling had dropped from £1 to US$4.86 in 1939, to US$4.03 in 1940 and US$2.8 in 1949. The U.S. dollar had replaced sterling as the most important global currency. The British government strictly controlled foreign exchange and the import of goods. Food was rationed, including eggs; Zhou was eating one, when he discovered it was actually yellow flour.

His stay in Britain coincided with the first period in office of a Labour government. Unexpectedly, in July 1945, it had won a landslide victory in a general election; the voters rejected Sir Winston Churchill, national hero of World War Two, in favour of a Labour Party promising sweeping social reforms. Most attractive was the introduction of the world's first medical free service. As a result, Britain had two medical systems, one private and one public.

In London, Zhou often visited the British Museum and marvelled at its collection of foreign exhibits, including treasures from Egypt and paintings of Dunhuang in western China. "I saw the British Museum and British Library and then knew that Britain was the centre of world culture."

In Britain, Zhou considered himself "left-wing", a result of his extensive reading and the influence of many Chinese intellectuals he knew. A majority of them opposed the Kuomintang government; some had joined the Communist Party, openly or in secret. Britain was the birthplace of Communism, where Karl Marx had written his most important works in the 19th century. Visits to the giant textile factories and housing estates of industrial Britain provided much of his raw material.

But, to his surprise, Zhou found that few in Britain respected Communism. He liked to buy *The Daily Worker*, a newspaper of the British Communist Party, but found it hard to find anyone selling it. It had no market among the British public. By contrast, during visits to the Continent, he found the Communist parties of France and Italy flourishing and popular. Both were active in the political process, as one of several parties, not like those in the Soviet Union and, later, China, where they were the single party. This fear of Communist Parties taking power in countries in Western Europe – and Japan – was a major reason for the lavish U.S. aid to both areas after World War Two.

During their visits to France and Italy, Zhou and his wife visited famous sites such as the Louvre in Paris and the ancient monuments of Rome. The latter gave him unexpected insights on language. In Rome, he saw large letters written on wall paintings; it took him some time to work out what they were – the names of Roman Emperors written in Latin. There were no gaps or punctuation between the words; only in later writing were these introduced. He also saw manuscripts written by William Shakespeare by hand; they were also hard to understand. The message which Zhou took from this was that spacing and punctuation greatly improved comprehension.

To Return or Not to Return?

During his final year in New York and his posting in London, Zhou was facing an existential question. The Communist Party was going to win the civil war in China and establish a new state modelled on the Soviet Union. Should he return to Shanghai and join it? Or should he make a new life for himself and his family in the United States? If he chose the latter, he was well qualified. He had fluent English, a strong academic record and a decade of experience in finance and teaching. This would enable him to find a good job in the U.S. His two years in New York had given him a wide network of contacts in the United States, in the business community and among the Chinese; many of them had chosen to stay and not return home, especially those working for or connected to the Kuomintang government.

While he had many friends in the Kuomintang government, he himself had not joined the party nor held an official post except for his time with the Agricultural Bureau in Chongqing. This gave him a liberty of choice not available to civil servants, especially senior ones; after a bitter civil war lasting four years and costing millions of lives, they knew they would not be well treated by the new government. Most escaped to Taiwan or the United States. In addition, Zhou's three years of working abroad had given him a salary in foreign exchange; his savings could pay for the start of a new life in the U.S.

From Chinese and foreigners, he received contradictory advice. On the Queen Elizabeth crossing the Atlantic, Americans told him not to go and live in a Communist state. Chinese travellers had a different opinion. One said: "Yes, life in the U.S. would be comfortable but without any meaning. It is already a developed country, so what contribution could you make as a foreigner? The U.S. has a deep well of talent. In addition, Chinese have no say in politics." They argued that, after 12 years of war and destruction, China had so much rebuilding to do, and qualified people like Zhou could make an important contribution. This was certainly the view of Lao She, the novelist he had met in New York. Like most Chinese intellectuals, he was on the left. One of his closest American friends, Pearl Buck, urged Lao She to stay in the United States; she had sponsored him to come and could help him settle there.

The winner of the Nobel Prize for Literature in 1938, Buck was a child of parents who were missionaries to China and spent about 40 years of her life there; few foreigners understood rural life there as well as she did. But Lao rejected her advice and returned home – with tragic consequences. After maltreatment and humiliation at the hands of the Red Guards, he took his own life by drowning in a lake in Beijing in August 1966.

Finally, after listening to all this different advice, Zhou decided to return to China. "At that time, not only China but also Europe was at a historical turning point. No-one could not consider the political questions. And no-one could say that he did not consider which side he should stand on politically. In such a situation, I took the decision – not go back to the U.S. but return to Shanghai!"

There were several factors in his decision. One was the fact that both he and his wife had large families well-established in China. Their relatives did not plan to leave. They were cultured and middle class and not the target of the revolution – large landowners, the wealthy and those working for or aligned with the Kuomintang government. Another was that, while Zhou spoke fluent English and had lived abroad for three years, he remained thoroughly a Chinese; he and his wife had sent their son to study in Suzhou, precisely because they wanted him to have a Chinese education in a Chinese setting. They could have sent him in an upmarket school in downtown Manhattan.

Had Zhou done undergraduate and graduate studies in the United States, he might have felt differently. "My mother was in China and did not want to move to the U.S. I did not want to be separated from her for a long period. At that time, many young people had hope for China after liberation. They wanted to return and do something for the country. Many people had this idealism."

Toward the end of his life, Zhou elaborated on this momentous decision. "I opposed the Kuomintang and supported the Communists because they advocated democracy. During the war in Chongqing, I took part in small-group meetings each month with Zhou Enlai. Every time he said that the Communist Party advocated democracy. We very much believed him and detested the dictatorship of Chiang Kaishek. In November 1940,

Chen Duxiu (one of the founders of the Communist Party) wrote that democracy of the proletarian was the same as that of the democracy of the capitalist class – a citizen's freedom of assembly, opinion, organisation, publication, to strike and organise opposition parties. After 1949, this important document could not be published."

To go home, he chose first to fly to Hong Kong, with British Overseas Airways Corporation. It offered an interesting itinerary – allowing passengers to get off the plane at several cities en route and explore them before getting back on the plane to continue the journey. This way Zhou and his wife were able to see places including Sicily, Karachi and Calcutta, where they had one memorable moment. The tourist car driving them came upon a cow crossing the road; it had to stop and give way to the sacred animal. It could not touch it. Such a scene was unimaginable in China or Europe.

In the spring of 1949, they finally arrived in Hong Kong and found it little developed. Fearful the new government would take over the colony, wealthy British people were taking their money and going home. Their decision turned out to be a mistake. The Xinhua Bank that Zhou worked for was nationalised by the new government in the mainland. But the branch in Hong Kong was a separate legal entity registered with the British government. So, it was not nationalised.

In Hong Kong, it was easy to buy an air ticket for Shanghai. Zhou was delighted to return. He was reunited with his mother, whom he had not seen for several years. His son had completed his studies in Suzhou and moved to Shanghai. Zhou went to the Xinhua Bank to report to his boss, Wang Zhishen. Shortly afterwards, Wang sent him back to Hong Kong to complete urgent business.

While he was there, Zhou received a secret telegram from a friend telling him not to return to Shanghai; but he did not give the reason. Zhou later learnt what it was. One of his missions in Hong Kong was to transfer the funds from Communist Party members in Shanghai. Under the rules of the Nationalist government, this was illegal. Nationalist officials in Shanghai had discovered the operation and put Zhou on a black list; if he returned, he would be arrested.

So, he had to wait in Hong Kong, along with many Communist Party members, until the People's Liberation Army took over Shanghai; it was only a matter of time. It captured the city on May 27, 1949. While Zhou was waiting, major Hong Kong newspapers commissioned him to write articles on finance and economy; most were published in the Sunday editions. The articles were collected and published in 1949 as a book, Zhou's first, with the title *Financial Questions of New China*. Published in Shanghai and Hong Kong, it had a big impact. Large libraries in the United States also ordered copies.

After the People's Liberation Army had taken over, Zhou was able to return to Shanghai. Since rail and air links had been cut, he travelled in a large British passenger boat hired by a Communist newspaper in Hong Kong. With him were many senior party members also stranded in the city. They arrived on June 3.

CHAPTER SIX

Creating Pinyin, Changing History

As he stepped off the boat at the port of Shanghai, Zhou was in a good mood. After 12 years of war, the country was finally at peace. The great work of reconstruction was about to start; he believed that, with the skills and knowledge he had acquired at home and abroad, he could make a big contribution.

The new government moved swiftly to improve the public order of Shanghai, rounding up criminal gangs, drug addicts and prostitutes. Residents felt safe as they had not done for many years; they even felt comfortable enough to leave their doors unlocked at night.

Zhou met an old friend who had just been appointed head of the Research Organisation at Fudan, one of Shanghai's most famous universities. The friend invited Zhou to become a researcher and teacher at Fudan; he was delighted to accept.

Since Zhou did not stick rigidly to the textbooks and included material from his life as a civil servant and banker, he became a popular teacher. One of his colleagues started a magazine *Economics Weekly*. Professors from Fudan worked together to edit it; Zhou wrote regular articles, devoting Saturday evenings and Sundays to the task. The magazine sold 10,000 copies each week, giving it considerable impact.

This was the research and academic atmosphere Zhou enjoyed. He also worked at Xinhua Bank, as secretary to the board. In one of his articles, he proposed the building of two major railway lines, one along the coast and the other along the Yangtze River from Shanghai to Chongqing and Chengdu. During World War Two, China's economy had fallen even further behind that of the West. The Sichuan basin was a closed economy, without a railway; full of resources, it should be opened up. China should build a national railway network before a national road network – that was the rule of economics, he said.

But this freedom did not last long. Things began to change rapidly, more rapidly than Zhou and his colleagues had imagined. For the new government, control of finance was a priority. It was one of the first sectors it nationalised, taking over both private and public banks. It set up the People's Bank of China to manage the banks. Since Shanghai had been the financial capital of China before 1949, it had more banks than any other city, as well as insurance firms, brokerages and other financial institutions.

By 1925, the gross assets held by Chinese-funded banks in Shanghai exceeded those held by foreign-funded ones. The 1920s and 1930s up to 1937 were the golden age for banking in the city. The shock for the thousands working in the financial system in the city was severe; what place, if any, did they have in the new order?

In October 1951, the Party launched a movement called "Three Against, Five Against". It aimed to combat waste, inefficiency and corruption. One main target was Shanghai's financial sector. Zhou found his bank had no business to do; instead, he went every morning at 08:00 to his office on the Bund for a daily meeting. The intensity of the meetings and the campaign provoked great fear among those attending; most had spent their careers in the financial industry and feared they could be targets.

No-one knew what to expect nor exactly what "crimes" the movement targeted. To escape "punishment", some took their own lives by jumping out of their offices on the Bund. Zhou described one meeting at which a deputy director was absent. After an hour, the news came that the man had arrived early that morning at the office and thrown himself out of his office. They found a body on the pavement, with the head smashed, making the person hard to identify. They went to his office and found a short suicide note saying: "I did not do bad things, only a little business and made a little money." Zhou became alarmed. "This made me very nervous. Often bankers jumped to their death from their office buildings. There were many suicides among those in financial and industrial circles. In fact, he (my colleague) was not a wealthy man. But, in the tense atmosphere of that time, he felt he had committed a crime."

After banking, similar radical changes arrived in education and the study of economics. As with banks, Shanghai had more universities than any other city in China. St. John's University was closed down, because it was an American missionary institution. Kwang Hua University was merged with other institutions to become East China Normal University. In 1952, the more than 20 faculties of economics at its universities were merged into a single institution, Shanghai Finance & Economics College. Each faculty had its own library; the merger of these libraries was chaotic.

Zhou was appointed chief of the research bureau of the new institution. He and his colleagues had to throw their old textbooks away; they had to use new ones, which were translations of Soviet ones. "The most important were political/economic textbooks. This was a faculty that had not existed before. There were no longer economics classes." The most important themes were the collectivisation of agriculture and the nationalisation of industry. The students became dissatisfied with the narrowness of the material. They told Zhou that they had read the textbooks and asked him to cover other material in the classes, such as Keynesianism; but he did not dare.

In a speech at Renmin University of China in Beijing, a Soviet economist had attacked the theories of British economist John Maynard Keynes; they became a taboo subject. The professors could not discuss or research freely as they had before. Zhou and his colleagues also lost the *Economic Weekly* magazine they had been writing; it was merged into a similar publication in Beijing.

In Zhou's view, China paid a heavy price for this ban on the study of economics. "The lack of development of the Soviet economy and that of China after 1949 had much to do with the lack of study of economics. Marxist economics is political economy and could not solve economic problems. It is the study of politics, not economics. Even today (1996-97), I fear, this problem can still not be discussed openly."

Later in life, Zhou was scathing about Marxist economics. "Marx denied the value produced by brain power and its contribution. But it accounts for 60-80 per cent of GDP in Western countries. In Japan, there are factories with no workers and, in the U.S., farms with no farmers (only machinery). In the U.S., some factories are 50 per cent owned by the workers. Are they exploiting or being exploited?"

In 2010, he said that the Soviet model, the blueprint for post-1949 China, had failed. "At its peak, 40 countries followed this model, now there are only six. The Soviet Union did not implement Socialism. [The Communist Party] overthrew one ruling class and then itself became the ruling class. The Soviet Union did the greatest harm to China. Once I saw a textbook of the Central Party School. It was laughable. On the outside, it said it was about Marxist economics. On the inside, there was much about (John Maynard) Keynes. This was progress." Zhou and his wife lived in a house off Shanyin Lu with a small yard; it was a quiet street in a residential neighbourhood. His son left his middle school in Suzhou and moved to one in Shanghai. The family lived in the house for five years.

A friend in Shanghai took Zhou aside and advised him not to hold two jobs, one in a bank and one at the finance and economics college. The bank salary was high by the standards of that time. He advised him to give up one of the jobs. Zhou wisely stopped working at the bank. It was a time to be low-key and stay out of trouble. "Between the winter of 1949 and 1955, there were constant movements and remakings, a real struggle to turn heaven and earth upside down. At that time, my awareness was very high. Although there were many things that I did not understand, I felt an enormous transformation and a great progress. So, I gave it my wholehearted support."

His wife was severely affected by the "Three Against, Five Against" movement. After 1949, she began teaching history at the most famous secondary school in Shanghai, Kwang Hua. Finding mistakes and inaccuracies in the textbooks, she proposed corrections and wrote to the *People's Daily*, the main Communist Party paper in Beijing. It published her proposals. They were sent to the People's Education Press; its president invited her to move to Beijing and work for his firm in the Chinese history department. Full of revolutionary fervour, she accepted the offer and moved to Beijing, leaving her husband behind in Shanghai.

The stated aims of the movement were to combat corruption, waste and inefficiency. Miss Zhang was guilty of none of them; but she was labelled a "big tiger" and branded a "counter-revolutionary". She had to write a "confession" of 20,000 characters; but it was not accepted. "I was terrified," she said.

Seriously ill, she returned to Shanghai for treatment. She told Zhou that she would not go back to Beijing. Finally, the investigators found nothing against her. This trauma profoundly affected her. She and Zhou decided that she should never return to full-time work again. Aware that such political movements had become part of "normal" life, they decided that she could not endure a second one and that it might cost her life. She said later: "So I calmly became a housewife for 40 years and never received a penny in wages from the state. As a very common person, I was very happy. If I had been working during the Cultural Revolution, I would certainly have died."

Life Transformed

In 1955, Zhou's life was transformed. The government moved him from Shanghai to Beijing to work on reform of the Chinese language. He had spent his university years and most of his professional life in Shanghai; it was the city where he was most comfortable. But he would stay in the capital for the rest of his life – 62 years. It was his work on the reform of the language that became his greatest contribution to China and the world.

The move may also have saved Zhou from prison or death. If he had stayed as a university professor of economics in Shanghai, he would have been a prime target in the Anti-Rightist Movement of 1957 and the Cultural Revolution, especially because he had worked in "imperialist countries" – the U.S. and U.K.

For the new Communist government, the battle against illiteracy was a priority. In 1949, China's literacy rate was about 20 per cent. On October 10, 1949, just nine days after the People's Republic was established, the China Character Reform Society was set up in Beijing. National leaders, including Mao Zedong and Zhou Enlai, took a personal interest in the issue. On October 6, 1954, the Standing Committee of the National People's Congress approved the Chinese Character Reform Commission (CCRC) as a government department under the State Council (the Cabinet). It appointed Wu Yuzhang as its director, with the status of

Minister. The new department started work on December 23, 1954. Its aim was not research but to change society. It had three missions – to simplify the characters, develop a system of Pinyin and promote the use of Mandarin (Putonghua) nationwide. In February 1955, the CCRC set up its Pinyin division.

In October that year, the CCRC and the Ministry of Education held a nine-day meeting in Beijing on Language Reform with delegates from all over the country; it was the biggest such meeting ever held in China. It invited Zhou to attend the meeting. Busy in Shanghai with teaching and other work, he asked how long the meeting would last and was told no more than a month. So, he took a month's leave from his duties and went to Beijing to attend. After it ended, the CCRC told him that he should stay in the capital and work for them.

The CCRC invited Zhou because, from his student days, he had a deep interest in how to use the Latin alphabet to write Chinese. He had written articles on the subject. In 1952, these were published in a book entitled *Chinese Pinyin and Character Research*. He did research and writing in his spare time, outside his regular work.

After 1949, the new government encouraged this kind of research. Those working on it in Shanghai published weekly and monthly magazines, to which Zhou contributed articles. In November 1954, he published in Shanghai a second book on the subject *The Subject of the Alphabets*, describing different systems around the world. All this had given Zhou a certain reputation in the field. As Zhou tells it, this book changed the mind of Mao – to accept a phonetic system based on the Roman alphabet, rather than one using strokes of the characters.

"At that time, the vast majority of experts preferred a system using character strokes, a minority wanted the use of Cyrillic letters and a small minority advocated Roman letters," said Zhou. "Chairman Mao preferred a system using character strokes. He asked me if I approved or not. I did not dare to oppose Chairman Mao, so I stayed silent. He asked me again and I did not answer. He understood and then called a recess. Later Hu Qiaomu (Mao's secretary) came to my house to ask my opinion. I said that I had written *The Subject of the Alphabets* and he took a copy for Chairman Mao to read. At our next meeting, Mao said that he favoured the Roman alphabet, in line with international practice."

History has shown this momentous decision to be correct. If Beijing had chosen a system based on characters or Cyrillic, it would not have been accepted by the international community; it would have been adopted only within China, and possibly the 12 countries whose language use the Cyrillic alphabet. China has much to thank Zhou for.

When the order came to work for the CCRC, Zhou insisted he was a layman in the field of linguistics and had pursued the subject only as a hobby. "I said that I could not (work for them)," Zhou said. "I was a layman in linguistics. And the college in Shanghai would not release me; it demanded that I return to work as soon as possible."

His lack of specialised knowledge of linguistics was only one reason why he did not wish to move to Beijing. A second was money. In Shanghai, he earned more than 600 yuan a month from three jobs, including economics professor; the new job in Beijing paid only 250 yuan a month. A third reason not to move was the weather. Beijing was cold, dry and windy, with fierce sandstorms; it had little rainfall. In addition, it was dirty, with rubbish piled on the side of roads. When the winds came, they blew the paper from the rubbish onto the faces of passers-by. Shanghai was warmer and more moist.

But Prime Minister Zhou Enlai personally issued an order to him to come. "That left me with no choice but to move to Beijing and work with the CCRC. I had no psychological preparation at all." The decision to assign Zhou to Beijing was a reminder of how the post-1949 system worked. It was the government, not the individual, who decided in what job and for how long a person worked.

His wife was not happy about the move. Their new home in Beijing was close to the office of the CCRC. When they arrived there, his wife discovered to her horror that it was next door to the People's Education Press, where she had recently worked and been treated so badly.

It had moved from the location of its previous office. She saw people she had worked with. She wept deeply in front of her husband and said: "I must leave and return to Shanghai!" The memories of persecution were too raw and too bitter. Zhou's colleagues at CCRC were supportive, saying that his wife had many talents to offer; she was offered a job at the CCRC or the Ministry of Culture. But, as we described above, she and Zhou had decided that she would not do full-time work.

Instead, she threw her energies into Kunqu Opera. In 1956, she and a friend ran the Beijing Kunqu Research Society; it organised performances of the opera and she performed onstage herself. This proved to be a very wise decision. Since she never belonged again to an official institution, she was sheltered from the savage campaigns of the next 20 years; with her sensitive personality, she might not have survived them.

One plus of their new home was its location, a five-minute walk from the Forbidden City and close to Jingshan Park in the centre of Beijing. It had formerly been the site of Peking University, a courtyard with trees, grass and flowers around a large square. In the mornings, Zhou used to walk to the Forbidden City, when few people went there, and wrote articles; it was quiet and the air was clean. Another plus was that he already knew several of his new colleagues and had a great interest in linguistics. These were compensations for his sudden departure from his colleagues and students at the Shanghai college and separation from his many relatives and friends who lived there. Although he had a lower salary, he had a car and driver assigned to him. Later, when the CCRC was downgraded to a bureau, he lost his driver and had to take the bus.

At the CCRC, Zhou was given an extremely important post. He was put in charge of the division responsible for creating a system of Pinyin. He was well suited to this job. He threw himself into understanding the different systems of Pinyin used around the world. "After I moved to Beijing, I could not continue with both economics and linguistics. The transfer was out of the blue. I had to put economics aside. I felt that, if you want to achieve something substantial and make a contribution to the country, you can only concentrate on one thing."

What to do with Chinese?

In 1605, Italian Jesuit missionary Matteo Ricci published a book that used Roman letters to write Chinese. It was the first Romanisation of the language. It was mainly intended for foreigners; Chinese paid no attention.

In the final decades of the Qing dynasty, there had been an intense debate in the government and intellectual community about how to make best use of the Chinese language. The two most urgent issues were how to improve literacy among the general public (especially in rural areas where it was below five per cent) and what was the best form for the written language.

In February 1913, two years after the Qing had been overthrown, the Ministry of Education of the new Republic of China held a Conference on the Unification of Pronunciation, with three aims; one was to establish a standard national pronunciation for the characters; another was to analyse the pronunciation in terms of their basic sounds; and the third was to adopt a set of phonetic symbols to represent these sounds.

After intense debate, the delegates decided against a system using Roman letters in favour of one using 39 phonetic symbols. They were created by a Chinese scholar named Zhang Binglin – also known as Zhang Taiyan – who based them on ancient characters; the modern readings contain the sound that each letter represents. They gave it the name Zhuyin Zimu, meaning phonetic alphabet, or Zhuyin Fuhao, meaning phonetic symbol; it is known today as bopomofo – and the principal system used in Taiwan.

Primary school students would first learn this alphabet and then the characters. The standard language was defined by fixing the pronunciation of 6,500 characters in a *Dictionary of National Pronunciation*. The delegates recommended that the Ministry of Education should promulgate the alphabet and that it be taught in schools. A major obstacle was that up to 25 per cent of the population, mainly in the south and east, spoke languages that were not Mandarin – such as Cantonese, Fujianese and Shanghainese; so, the sounds they read would not be familiar to them.

Over the next 30 years, this alphabet was promoted by reformers and in some places, but it was not adopted nationwide because of a lack of consensus. Many said that a Romanised alphabet would be better. Some even suggested that characters be abandoned completely and replaced by Esperanto, an international language invented by Ludwik Leyzer Zamenhof, a Warsaw-based ophthalmologist in 1887; it became the world's most widely used newly constructed language.

One main proponent of a romanised alphabet was Chao Yuenren, whom Zhou had met in the summer of 1946 at the University of Michigan. His proposal used only the 26 letters of the Roman alphabet and was detailed and well-argued. The use of this alphabet became widely accepted among language experts in China. While he was pursuing his career in finance and business, Zhou followed all these debates with great interest. He especially liked the proposal of Chao Yuenren.

The Nationalist government supported the promotion of the national language and the use of a romanised or phonetic system to help people learn it. But, from its formation in 1925 until its retreat to Taiwan in 1949, it had more urgent priorities than language reform – unifying the country and establishing a new system of government, fighting the Japanese and also fighting the Communists.

In addition, in the early 1920s, the government carried out a major change. It accepted the proposal of many reformers to discard Classical Chinese as the standard written form in favour of the vernacular language. This made reading and writing accessible to millions of people. Learning the Classical version, the official written form used for many centuries, took years of intense study. Reformers argued that the vast majority of people had no opportunity to do this and that those years could be better spent learning mathematics, science, English and other subjects needed to turn China into a modern nation and catch up with Japan, the United States and Europe. This switch to the vernacular was a historic reform of the language.

Learn from the Soviet Union

For the Chinese Communist Party born in 1921, the Soviet Union was the model. One year after the Bolsheviks took power in 1917, they carried out a major reform of the language by eliminating four letters, to make the orthography simpler; they retained the Cyrillic alphabet and did not adopt the Roman one.

Among those witnessing the revolutionary changes in Moscow was Qu Qiubai, a Communist party member and Russian-speaking journalist for the *Morning News of Beijing*. He visited the home of Leo Tolstoy at Yasnaya Polyana with Sofya, granddaughter of the great man; saw Vladimir Lenin address a group of Communist delegates; and witnessed the funeral of Pyotr Kropotkin. Lenin said: "Latinisation is the great revolution of the East."

Qu worked with linguists at the Soviet Oriental Institute of Leningrad who specialised in Chinese to develop an alphabetic script for Mandarin. They called it xin wenzi – literally meaning "New Characters" and usually translated as Latinised Characters. They produced textbooks, newspapers and other reading material in the new language. The script was easy to learn. At the first national conference on the new system, in Vladivostok in September 1931, the authors declared that the characters were a product of "an ancient and feudal society and had become one of the tools of the ruling class to suppress the toiling masses, an obstacle to mass literacy and unsuitable for the modern age."

When the new system was brought to China, it was welcomed by left-wing people and promoted by the Communist Party in areas it controlled. In 1936, Mao Zedong told American journalist Edgar Snow: "In order to hasten the liquidation of illiteracy here, we have begun experimenting with xin wenzi – Latinised Chinese. It is now used in our Party school, the Red Academy, the Red Army and in a special section of the Red China Daily News. We believe Latinisation is a good instrument with which to overcome illiteracy. Chinese characters are so difficult to learn that even the best system of rudimentary characters does not equip the people with a really rich and efficient vocabulary. Sooner or later, we believe, we will have to abandon characters altogether if we are to create a new social culture in which the masses fully participate."

At press conferences in Yenan, Mao noted how foreign journalists took notes far faster than the Chinese reporters. Over 30 journals appeared in xin wenzi, in addition to many translations of foreign works, modern Chinese literature and textbooks.

After its victory in 1949, the new government had achieved the stability necessary to implement wide-ranging language reform. But Mao, and his Prime Minister Zhou Enlai, encountered intense opposition to abolishing the characters; educated people regarded them as a national treasure passed on by generations of scholars through thousands of years of history. They were unique to China. The new government badly needed the help of these people to reconstruct the country, especially its education system. So, Mao gave up the idea of doing away with the characters completely; but he remained determined to carry out far-reaching reform.

In December 1949, Mao Zedong went to Moscow for two meetings with Soviet leader Josef Stalin. He asked him: "What should we do about language reform in China?" Stalin replied: "You are a great country and should have your own alphabet and not simply use the Latin alphabet characters." After his return to Beijing, Mao called for minzu xingshi – "national form writing system". Over the next three years, scholars did intense work and came up with four different proposals. But the CCRC rejected all of them. Its director, Wu Yuzhang, reported this to Mao and said that the use of the Roman alphabet was the best option. Mao agreed.

Then a new threat loomed. After initially promoting the use of the Roman alphabet, the Soviet Union had turned against it. Instead, it wanted a "Slav alphabet", using Cyrillic, the script in which Russian, and several other languages, are written. Soviet experts came to China and urged the government not to use Roman letters. A Soviet Vice Education Minister arrived in Beijing and met Vice Premier Chen Yi; he made the same proposal, saying that such a choice would deepen the ties between the two allies.

This was a critical moment. The new government was not 10 years ago; the Soviet Union was its biggest ally and the model for the state it was building. The Soviet Union had sent more than 10,000 technical specialists to work on dozens of major industrial projects in China; Russian was the first foreign language for Chinese to study in school and thousands of Chinese were studying in the Soviet Union. The pressure to follow the Soviet example was strong. But Chen held his nerve. He rejected the idea, saying that China needed to connect with the overseas Chinese community in Southeast Asia and around the world; they knew Roman letters but not Cyrillic. For Zhou and his colleagues, this decision was a great relief.

Creating Pinyin

The challenge facing Zhou and his team of less than 20 people was daunting. The strongest argument for retaining the characters was that there are so many homonyms with the same sound. The Chinese-English dictionary I use, published by the Beijing Foreign Studies University in 1978, has 1,750 pages – it is modest in size. There are a total of 70,000 characters; a person with a university education knows 5,000-6,000 and with a secondary school education 3,000-4,000. For daily life, knowledge of at least 2,000 is needed. Take "yi" in my dictionary – 31 pages and 117 characters with the same sound!

Zhou and his colleagues wanted to devise a system for transcribing the sounds of Chinese characters into the Roman alphabet; it would be called Hanyu Pinyin, Chinese phonetics. Since the late 19th century, the English-speaking world had used a form of romanisation called Wade-Giles, named after Thomas Francis Wade and Herbert Giles, two British scholars who completed it in 1892. In Taiwan, it remains a form widely used today. Zhou believed he could do better.

Zhou and his colleagues produced the first draft of Pinyin in 1955. The next year they came up with a modified draft, with two readings for each character. This was rejected as too complicated and they returned to the idea of one sound per character. But the 26 letters of the Roman alphabet were not enough, so they added six more. But this plan met strong opposition, including from the Ministry of Posts and Telecommunications; it said that this would make telegrams impossible to send and that foreign post offices would be unable to read the six new letters.

So they went back to 26 letters. Zhou worked intensively for three years: "We received more than 4,000 letters from people in China and overseas. They were full of questions, requests and suggestions. I had to reply to everyone in academic language." He said that there was great resistance to basing Pinyin on the Roman alphabet. Some called him a "slave of the West". "I insisted on it. Perhaps it was due to my time overseas. I always envisaged Pinyin being useful to foreigners too. I see it as a bridge between China and the rest of the world, a bridge between cultures."

He said that it worked better than previous phonetic alphabets used in China, like Zhuyin Fuhao, and Wade-Giles. "It exists to help and supplement Chinese characters, not replace them. If that were not official policy, people would oppose its use all together. Chinese characters have been used for several thousand years and have deep roots. Their development has been over a long time. Replacing them would not be done in a few days, not in a hundred years. They asked me how many hundred years. I replied: 'In 500 years' time, ask me again.' "

Zhou and his colleagues completed the final version in 1958. They created an alphabet as close as possible to the sound of the characters; they added to each one of the four tones used in Mandarin. It uses 24 of the 26 letters of the Roman alphabet, but not "i" or "u". Peking in Wade-Giles became Beijing in Pinyin, Mao Tsetung became Mao Zedong.

Mao gave his approval. Then Pinyin was formally approved by the CCRC and the National People's Congress, the Parliament. "After the NPC passed Pinyin, China had its own standard. Spending those three years was very worthwhile," said Zhou. "Many practical problems were well resolved. The remaining problems were very hard to solve. Why is that? Because this phonetic alphabet was not designed for Chinese."

On January 10, 1958, Prime Minister Zhou Enlai declared that the three objectives of language reform were simplification of the characters, spreading Pinyin and spreading Putonghua as the national language. Since 1958, these objectives have remained unchanged.

Spreading Pinyin

From the autumn of 1958, the government began to spread the use of Pinyin nationwide. The first place was the primary school, starting in the first year for children of seven years old. They studied Pinyin before they learnt the characters, because it was easier. Publishers produced school textbooks with Pinyin above each character.

With a colleague, Zhou travelled to many parts of the country to explain and popularise the new system. In the late 1950s, the transport system in the southwest was undeveloped, with few railways; travelling by air was the quickest way. At that time, not many people in China travelled by plane, especially away from major eastern cities like Beijing, Shanghai and Guangzhou. Zhou and his colleague flew first to Xi'an and stayed in a giant hotel built to receive Chairman Mao – but he never stayed in it. There were only a handful of guests. They were due to fly to Chengdu, in the southwest, but the weather was poor. The airlines were very cautious; they did not fly after dark nor if the weather was bad. So, passengers were very safe. This gave the two men several free days in a high-class hotel.

At the airport, the staff informed them that there were only a handful of passengers for the flight to Chengdu, and not enough weight on the plane. So, they had to add blocks of stone in the storage area, to make it safe to fly – like ballast in ships. The staff joked with Zhou: "Three aeroplanes would be enough for China." During these tours, Zhou and his colleague explained Pinyin to teachers and intellectuals; they responded favourably. He also gave a series of lectures at Peking University and Renmin University of China. He turned the lectures into a book *An Introduction to Reform of Chinese Characters*. The first edition came out in 1961 and the second in 1964; a Taiwan company also published it. This book had an enormous impact, among leading officials and the public. Of Zhou's 49 books, it was the one that had the greatest influence. It was a subject that interested both Chairman Mao and Prime Minister Zhou Enlai.

In 1985, after three years of intense work by Japanese scholars, the book was also published in Japanese. The next year, Zhou was invited by a major Japanese publisher to Tokyo to take part in an academic conference on Chinese characters attended by 500 guests. It was to mark its publication of a Japanese-Chinese dictionary. There were guests from countries which had in history used Chinese characters – North and South Korea and Vietnam, as well as China and Japan.

Hangul

Now let us look at another country that formerly used Chinese characters as its written language and took a different course – Korea. In the 15th century, King Sejong the Great faced the same dilemma as the leaders of China in the 20th century – the vast majority of his people could not read the written language, because it was too difficult and complicated to learn. So, like Mao and Zhou, he entrusted a group of scholars to devise a new way to read the language.

They did not use Roman letters but invented new symbols designed to reflect as closely as possible how Koreans spoke. This was Hangul, with 14 consonant letters and 10 vowel letters, created in 1443. Like Zhou, the King intended it to complement, not replace, the characters. For the next 450 years, the two scripts co-existed with the court and educated class using the characters; they remained the official language.

During the Japanese occupation from 1905-45, there was a resurgence of Hangul as a symbol of Korean identity – even though the Japanese banned Koreans from using it; over 100 newspapers were published in the language. After the Japanese surrender in 1945, the new state of North Korea banned the characters; since then, it has used only Hangul. South Korea continued to use both. Students had one class a week learning characters; about half of the articles in newspapers used them. In 1993, this changed. The newspapers became 100 per cent Hangul, and it was unnecessary to learn the characters. Hangul is one reason for South Korea's prowess in high technology; it is easily adapted to the computer and the Internet.

One Korean official said: "Hangul was developed by King Sejong working with experts. His purpose was to enable common people to read and write something they could understand. They could not read Chinese characters, which were only for the elite and highly educated. The language is very scientific; it is regular and logical, with no exceptions, no male or female; for past, present and future, you simply add an additional word. You do not need to memorise it. It was a completely new language; those who devised it used sounds imitating the tongue and palate. You can mix and match the words.

"Hangul is more supple than characters and one reason for the boom in Korean culture. It makes it easier to communicate, to write and speak. For the Internet, it is very convenient – fast to type and send messages. It is one reason why South Korea has developed so fast in science and technology and in the digital sphere. Hangul adapts easily to the computer," the official said.

Those who wish to read pre-1945 documents in the original or material from the Chinese-language world must learn the characters. "But many of the pre-1945 documents have been translated into Hangul. Then there is no need to learn the characters," he said. "We Koreans are very proud of Hangul."

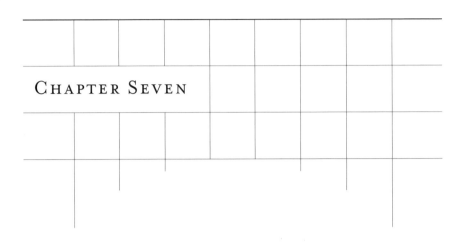

CHAPTER SEVEN

Saved From Prison By Linguistics: Millions Starve To Death

Zhou and his wife adapted quickly to life in Beijing. Three of her sisters and three of her brothers lived in the city, so they had a wide family network. In addition, they had many friends living there.

After her decision to stay away from paid employment, Madame Zhou threw her many talents into the study and research of her beloved Kunqu; she became director of the Beijing Kunqu Research and Study Association. They had many friends among performers and enthusiasts; they provided free tickets to performances. So, the main entertainment of the weekend was attending a performance; these sometimes went on so late that the couple arrived home after midnight.

Fortunately, Zhou had a driver allocated by the CCRC; the man graciously waited for the performances to end to drive the couple home. Many of the nation's elite were aficionados too. At the performances, they often saw former Emperor Pu Yi, Prime Minister Zhou Enlai and Kang Sheng, a senior Party leader. Zhou himself was not a great lover of Kunqu but went to accompany his wife. "I liked Western music. As a young man, I learnt the violin. When I went to listen to Western music, my wife came with me. When she went to listen to Kunqu, I went with her."

In 1959, their first grandchild was born, a girl named Heqing, to son Zhou Xiaoping and his wife. She was named in honour of the daughter they had lost in Chongqing – her name was Xiaohe and she passed away in Chongqing. At that time, Xiaoping was studying at Peking University, where tuition and food were free. Shortly after the baby was born, he went to study in the Soviet Union. The baby did not want to sleep with her nanny; she preferred to sleep with Grandmother. So Zhou was kicked out of the marital bed and slept in a single bed below a window.

The move to Beijing resulted in a sharp loss in income. They had only

Mid-1950s, Zhang Yunhe on the stage of Le Yi Girls' High School performed Sifan under the direction of Zhang Chuanfang.

one income, that of Zhou, which was 241 yuan a month. Since his wife did not work, she had no income nor health insurance. They paid Zhou's mother a monthly allowance of 90 yuan; in her final years, she required expensive medical treatment and drugs, many imported, amounting to 700-800 yuan a month.

"For the first time in my life, I went into debt," said Madame Zhou. "The amount rose to 4,000 yuan. The pressure became like an illness. I could be listless throughout the day." In 1964, Zhou's mother passed away, aged 96. She had lived a traumatic life; but, in her later years, she had enjoyed peace and comfort and seen the birth of her great-granddaughter. By 1966, Madame Zhou had been able to pay back all the debts, except 200 yuan borrowed from a cousin.

Zhou was invited to join the Chinese People's Political Consultative Conference (CPPCC), the top advisory body to the government. He served in it for more than 20 years. This brought status and privileges, as we will see shortly. Through his participation, Zhou met national leaders, including Mao Zedong, Zhou Enlai and Deng Xiaoping.

Saved from "Anti-Rightist" Campaign

The next political campaign was the "Anti-Rightist" movement launched by Mao in the summer of 1957. The main target was intellectuals suspected of "rightist" or "capitalist" opinions or opposing the Party. Lasting for two years, it caused the persecution of more than half a million people. Work-units were given a quota of five per cent of "rightists" whom they had to uncover, dismiss and punish. Often there were, in reality, no "rightists".

The CCRC was a new institution and not a primary target; linguistics was not ideological. Nonetheless, it had to "unearth" the five per cent of staff who were "rightists". It designated several young employees. Zhou described the judgement as "very tragic"; he himself was not affected. The next impact of the movement was to close two research departments in CCRC; Zhou was head of one of them. He was little affected by this decision and continued to work on his own.

The movement also involved making reports on individual staff about their behaviour and thinking. Fortunately, that on Zhou was favourable; it said that he accepted the movement and caused no trouble. His strategy was to concentrate on his work and research and avoid meetings where you had to express an opinion and left yourself vulnerable to criticism for what you said. He kept a low profile and was careful what he said in public. "During the 10 years from 1956 to the start of the Cultural Revolution in 1966, my own life was quite stable and I was not drawn into the many movements," he wrote.

But he was touched by the disgrace of Ma Yinchu, president of Peking University and one of China's most distinguished economists. After studying at Yale and Columbia Universities in the United States, he returned to China. He taught at universities in Shanghai; Zhou heard many of his lectures. After 1949, the two men worked together on articles for *Economics Weekly* magazine and became friends. In June 1957, Ma presented to the National People's Congress, the Parliament, his "New Population Theory".

After the establishment of peace in 1949, the birth rate had soared. Analysing the figures, Ma concluded that family planning was essential for China's development and the country could not sustain so rapid a growth in the population. He is considered the father of family planning in China.

But Mao dismissed his ideas as "anti-Socialist"; he believed a larger population made the country and its army stronger and more powerful. Mao had Ma removed from his post and the media banned from reporting his proposals. Zhou received a notice telling him to take part in a "mass criticism session" of Ma attended by 10,000 people. He did not want to go. He said that, as he was no longer an economist but worked in linguistics, the matter did not concern him; he did not attend.

The disgrace cost Ma 22 years of his life. It was only in July 1979 that the Central Committee of the Party formally apologised to him, saying that events had proved him right. In 1975, the population reached 916 million, an increase of 67 per cent over 547 million in 1950. In September 1979, Ma was made honorary president of Peking University, at the age of 97. He died in May 1982. Official estimates said that, had the government followed Ma's sage advice, it would have saved 300 million "unwanted" births. In 1980, the government implemented the "One-Child Policy" for the vast majority of families, the most stringent population control in human history; it lasted until 2016.

While the "anti-rightist" campaign did not touch Zhou directly, it did not spare many of his friends. One was Zhang Naiqi, Minister of Food from 1952 to 1957. Zhou had known him in Shanghai during the 1930s; educated and articulate, he was well-known for his left-wing views. But, in 1957, he was designated a "rightist", dismissed from his posts in June of that year and persecuted. His many friends, especially in Shanghai, found this unthinkable. Zhang disappeared from public view; there was no news about him.

One day a friend gave Zhou an address for Zhang, in an apartment building in an eastern suburb of Beijing. He went to the building; no-one there had heard of Zhang, but Zhou persisted. Finally, on the top floor, the eighth, he knocked at a door, and Zhang opened it. It took a few seconds for the two to recognise each other. Zhang was not in a good state. The room had a large bed, a toilet and an old sofa covered with clothes. Zhang cleared the clothes, so that Zhou could sit down. They talked for about 30 minutes. "It was very strange," wrote Zhou. "Outside there was a rumour that, within the Party, he had been rehabilitated, but he did not know." The persecution ended Zhang's professional life. In May 1977, he died of illness in a Beijing hospital.

Zhou's move to Beijing turned out to be a godsend. A primary target of the Anti-Rightist movement were scholars of the economy, especially those trained in the West. "Marxism did not need economists. It only wanted political economists. Economists knew the defects of Communism." They were fired from their jobs, criticised and sent to do labour in the countryside. Shen Zhiyuan, head of research at the Shanghai Economics Research Institute and a close friend of Zhou, took his own life, as did one of his Ph.D. students.

"Many economists in Shanghai did not say one word wrong but were branded rightists. As a professor, you had to write articles. One article could cost you 20 years in prison. Many were sent for reform through labour," said Zhou. "When they were rehabilitated, they were old and sick and unable to work."

But, fortunately, linguistics was not a target of the campaign; it was not considered sensitive. So Zhou was left untouched. If he had stayed as a professor of economics in Shanghai, it would probably have been a different story. "Friends told me that, if I had not gone to Beijing to work in the CCRC, I would certainly have spent 20 years in prison. Economists in Shanghai were a major target (of the movement), but not those studying linguistics. I escaped a disaster." During his life, because of his foreign language skills, Zhou was several times invited to join the Foreign Ministry. His wife fervently – and correctly – opposed this idea: entering politics meant certain trouble.

The campaigns kept coming. In 1958, Mao launched one against "The Four Pests" – rats, mosquitoes, flies and sparrows. With his colleagues, Zhou was ordered not to go to the office but to take part in the campaign. They spread poison on roofs, trees and the ground; in the evenings, they shook the trees, to force the sparrows to fly everywhere until they dropped dead from exhaustion. Among the collateral damage was a small bird that lived on a tree outside the window of Zhou's home. He lived in half of a Western-style house built in 1912 for a German professor at Peking University. Each morning Zhou heard the sound of the bird as he was getting up, a pleasing morning call. But the bird died in the cull of the sparrows. And the tree also died, for lack of maintenance.

"Biggest Disaster in the History of China – 45 Million Dead"

In 1958, Mao launched his next campaign, the Great Leap Forward (GLF). One objective was to double China's steel production, to attain the level of the Soviet Union by 1960 and overtake that of the United Kingdom. Zhou and his colleagues were ordered not to go to the office but smelt steel, using any metal they could find, including keys, window frames and other items in the house. On business trips by train to Tianjin and Shanghai, Zhou saw the lights of small and large furnaces at night and the stumps of trees along the lines cut down to provide fuel for them. "This was a very laughable project and did great harm to China," he wrote. The project was later abandoned because the steel produced by peasants was low grade and unusable.

Another element of the GLF was to follow the Stalinist model and collectivise agriculture. "Farmers are very conservative and did not want collectivisation. They regard their land as their baby. If you ask them to hand it over, they definitely do not agree," Zhou wrote. This turned out to be the most disastrous mistake of Chairman Mao's rule. Farmers were forced to live in "people's communes" and eat in communal canteens; they could no longer own or rent their own land and had no incentive to increase production. In addition, many were mobilised for mass projects, like backyard steel production and building roads, dams and irrigation schemes. As a result, grain production fell. But the official media did not say this, nor was it included in internal reports within the government; no rural official dared to admit a fall in grain output in his area.

Living in the city, Zhou felt the impact in the falling supply of food, including grain and vegetables. "Our family could not buy enough to fill our stomachs," he wrote. But he had a privilege denied to the vast majority of his fellow citizens – he was a member of the CPPCC; it had a club with a restaurant where he could go with his wife. The two went there every day for a meal; as a result, their limited food at home would last longer. They usually ate the western menu, which cost eight yuan, a large sum at that time, and often sat next to Pu Yi, the last emperor of China.

Even better, as a CPPCC member, Zhou had a pass which entitled him to go to a special store. It was on the second floor of a food market; he entered by the back door. This store had many products, including meat, sugar, eggs and tobacco not available to the general public; each CPPCC member could buy a fixed ration of each. "Cigarettes were a big item. There was a severe shortage. I did not smoke, but one of my friends was an addict. He smoked from morning until night. He could go without food but not without cigarettes. So, I bought them during every visit and saved him from 'tobacco hunger.'"

These stores were a feature of Socialist states. They were open to senior government and Party officials but not the general public. Once I was talking about them with a Beijing friend. I asked him if they were appropriate in a Socialist state that claimed to be "egalitarian". "Western countries are no different," he replied. "Britain has up-market stores like Harrods and Fortnum & Mason. They are only for the rich; ordinary people do not enter them. All societies are unequal."

Gradually, Zhou came to realise the extent of the disaster that had befallen his country. His family employed a domestic maid who had worked with them for many years; she came from Anhui, a poor agricultural province in eastern China. Her husband and daughter, who had stayed in the family's home village, died of hunger. "Later I came to learn that very many people had died of hunger in Anhui. The figure that many people told me was about 5.5 million."

A friend who worked as a reporter for the *People's Daily* was sent to Gansu province in northwest China to research the people's communes. He told Zhou that, in mountainous areas, he found dead bodies, frozen and naked, after they had been stripped of their clothes. He saw many such frozen corpses. "After he returned to Beijing, he said that he did not dare to report these things." Rather than speak about them, it was better to take sick leave and go home.

Zhou's next moment of truth came several years later in a railway carriage after a trip to the Great Wall; he and his friend Xu Qian'an had visited a memorial hall honouring the father of China's railway system Zhan Tianyou. It was a very slow train. With Zhou and his friend were workers from a coal mine; migrants from Anhui, they started to chat. One miner asked another: "At that time, how did you not die of hunger?" The man replied that he was a cook. Each day he put a little food into a box, which he buried in a hole in the ground. In the middle of the night, he told his boss and the two ate the contents of the box. The others said that the hunger reached the level that men gave their daughters to other men. Another said: "Two men opened the cover of a large pot and found the dead body of a girl inside. The two men collapsed and died. They were already almost dead and wanted to open the pot to find something to eat. The shock of what they saw killed them." Zhou said that, the younger you were, the more likely you were to die, because you needed more nutrition. "There was no end to telling these tragedies. These men told these stories to each other and we heard with our own ears. We got off the train in Beijing and could never forget this experience."

Zhou said that these campaigns drove China to bankruptcy. "Many people say that the Cultural Revolution drove China to the brink of bankruptcy. This is incorrect. It was bankrupt long before that. I was in Beijing, a member of the nomenklatura, but there were no peanut kernels to eat. Grain coupons did not give us enough to eat."

China did not abolish the people's communes until after the death of Chairman Mao in September 1976. As the country opened up, Zhou was able to read foreign material. He learnt that, in the Soviet Union, collectivisation of agriculture had also been a disaster, leading to the death of millions of farmers, especially in Ukraine. So, the Soviet government switched instead to collective farms. The Communist countries of Eastern Europe controlled by the Soviet Union after 1945 did not enforce collectivization of farming. So Mao was pursuing a policy that had already been abandoned by most of the Socialist world.

Looking back on these movements through which he had lived, Zhou made this judgement: "The anti-rightist movement was a very big mistake. But its influence was at that time not too large, since it impacted only a portion of society. It did not influence industry or agriculture. Of course, the movement was against knowledge. Its long-term impact is hard to estimate. It was a very important reason for China's backwardness." He said that the biggest impact on China was that of the people's communes. "This led to the death of tens of millions of farmers and a very sharp fall in output of grain. The people had nothing to eat … In recent years, I read foreign materials which gave a conservative estimate of 45 million dead from starvation. It was the biggest disaster in the history of China."

CHAPTER EIGHT

"Cow Pen" And Blanket Bombing By Wild Geese

In 1966, Chairman Mao launched the Great Proletarian Cultural Revolution (GPCR). It was the political campaign that had the greatest impact on Zhou during his life. Mao started it because he believed the revolution was losing energy because of conservatism and lethargy within the party. He also thought many of his colleagues regarded him as too old – he turned 73 that year – and wanted to sideline him from making policy and become "a Buddha on a shelf".

At the end of August, Mao stood on top of the gate overlooking Tiananmen Square and reviewed tens of thousands of screaming Red Guards, waving copies of his book of quotations. The government closed China's schools and universities and encouraged the Red Guards to attack their teachers, parents, school administrators and Party leaders. Zhou was 60 and working in the CCRC; unlike history or ideology, language reform was not a sensitive subject. Most people, including Red Guards, did not know what it was.

Zhou and his colleagues immediately felt the impact of the new campaign. They continued to go to the CCRC – not to work but to study political documents, including editorials in the *People's Daily* and the writings of Mao. For the first two years, he was able to go home early in the afternoon and rest at home. Raging bands of Red Guards enforced the new order; they were subject to no restraint or control. The atmosphere was terrifying; intellectuals like Zhou were a common target of attack.

During the third year, institutions started to build "cow pens", to detain "undesirable" people. The CCRC had three large garages for its cars; from one of these, the cars were removed and replaced with beds, to turn it into a "cow pen". Two undesirable groups were "capitalists" and "counter-revolutionary academic authorities". Zhou was designated as one of the latter.

"They did not hold a special meeting to criticise me. But there was a general criticism and struggle meeting." Red Guards carried a banner on the street with his name and insults written next to it; on a trip to the zoo with her mother and grandmother, young Heqing saw her grandfather's name but, fortunately, did not understand the insults.

"They (the Red Guards) came to search our home. We should consider them well-mannered. They took some books but did not damage the home. I was privileged. They did not beat me. One 'counter-revolutionary' told me that, on the investigation, they had found no links between me and foreign countries. Some of my colleagues were beaten to the point of being unconscious. My wife could not bear it."

In September 1968, Zhou had to move into the "cow pen", along with two other senior staff of the CCRC and 20 of the other 70 employees. They did manual work, such as cleaning the floors, collecting the rubbish and weeding; it was not too harsh. They read the works of Chairman Mao and had to memorise them; they were not allowed to take notes and had to hand the books back at the end of the classes. At meals in the work canteen, those from the "cow pen" had to wait at the back of the queue, to allow everyone else to take their food first. Those in the "pen" were allowed to go home on Saturdays and spend one night away before returning the next day. Zhou's monthly salary was cut to 35 yuan; after paying fees and bills, he did not have enough to live on.

Burning Books and Photographs

Zhou was also affected by another part of the GPCR – "destroying the four olds"; these were pre-1949 elements of Chinese life – Old Ideas, Old Culture, Old Customs, and Old Habits. This included the burning of old books. "This created a terrifying atmosphere, making people get rid of old things in their home, including diplomas, documents, university appointment letters and photographs, even old furniture. You could not have a foreign exchange, silver or gold."

Zhou had stored many of his books in a small room in the CCRC; but, as a "black element", he was not allowed to go there. So, he and his wife had no choice but to sell their books to second-hand bookshops, for a fraction of their true value. His wife had many precious photographs, taken with her parents when she was a child and later in life, as well as those of Kunqu performances. She had no alternative but to destroy them. "The atmosphere became more and more tense. Some did not implement the 'destroying the four olds' sufficiently and became the object of political struggle. This became more and more fierce."

The enforcers of the policy from the CCRC came to Zhou's house to inspect his books – but only the Chinese ones. They did not touch those in English and other foreign languages – presumably because they did not know what they were; so, he was allowed to keep them. His friends and relatives received similar treatment. One of the worst affected was Shen Congwen, his brother-in-law and one of China's most famous novelists. He had an enormous collection of books. Red Guards removed all of them and forced Shen and his family to live in a small room of his house where he had formerly stored books.

"At that time the whole country stopped work, apart from farmers tilling the land and workers going to factories. With the exception of a very few places, all schools, research institutes and government offices stopped working. That had never happened in the world. The powerful central government issued an order to the whole country not to work. This was without parallel in history," Zhou commented.

Despite all these restrictions, Zhou remained intellectually curious. Having lost most of his books, he went to the nearby Beijing Public Library to borrow new ones. The staff there told him that he could only borrow books related to his specialty, but not other kinds. He asked to borrow *The Merchant of Venice* by William Shakespeare. "No," said the librarian. "That is literature and you work in linguistics." The book was classified under the category "Basic English".

Zhou said that he was researching basic Chinese and needed to compare basic English. After a long discussion and out of politeness to an elderly scholar, the librarian agreed to lend the book. Zhou saw a continuum from the "anti-rightist movement" to the Cultural Revolution. It was a campaign against intellectuals; it did not allow them to write articles or read books, ancient or foreign. The CCRC and similar institutions were targets.

Among Zhou's friends, there were many tragedies. In the evenings, he and his wife used to go for walks with two friends in the nearby Jingshan Park. One was an engineer who had been educated in Europe and amazed them with his encyclopaedic knowledge; he had learnt much of it from reading books in German. "Our evening talks in the park were very happy," he wrote. "We did not talk about politics or the Cultural Revolution. Around us, things were becoming more and more terrifying. We did not speak of it, only of interesting things."

But such a pleasure could not last long. Red Guards went to the house of the engineer and took him away. A few days later, Zhou learnt that he had died. He suffered, it turned out, from serious diabetes, which required regular medicine and a strict diet. The Red Guards locked him in a small room and beat and criticised him. Without proper medication, he quickly died. Red Guards also went to the home of their other walking companion, an educated lady; they stole everything of value and smashed many of her belongings. They themselves moved into her house and put the lady and her family into a small room that had been used to store firewood.

The lady had a niece of secondary school age, also a member of a "bad" family. How could she improve her status? One way was to move to the countryside and do manual labour. In so doing, she showed she was no longer a "counter-revolutionary". Initially, she was happy to do this; she married a farmer and they had a child. But, after the end of the Cultural Revolution in 1976, the couple fell out. She was a city girl from a cultivated family and her husband a peasant with no education; they had little in common. The marriage broke down with much bitterness; they divorced.

For another educated girl of a similar background, it was a happier ending. Initially, she refused to go to the countryside; but she was forced to settle in Heilongjiang province, in the far north on the border with the Soviet Far East. It is the coldest region of China. Looking ahead to an end of the Cultural Revolution, her grandfather advised her that under no circumstances should she marry anyone in the village. Overcoming every difficulty, she followed his advice. After 1976, she was able to return to Beijing and resume the life she had before. She could choose the husband she wished.

In this chaos, the law was made by roving gangs of Red Guards. To try to protect itself, the CCRC organised its own team of Red Guards. One of their most popular weapons was big-character posters; they filled the walls and the sides of buildings. Some named Zhou as a "bad element". Zhou's wife became so frightened that she and her granddaughter locked their home and went to live in the home of her son and daughter-in-law;

he was a scientist at the Chinese Academy of Science and lived in the Zhongguancun district of northwest Beijing. He and his wife had been sent to a "May Seventh Cadre School" in Hubei in central China. Zhou's wife's decision not to take up full-time work proved a very wise one. Her status as a "housewife" had given her a measure of protection; she had no work-unit whose Red Guards investigated its employees.

Zhou and the other "bad elements" had to attend the meetings at which the Red Guards criticised people like him; there was severe verbal and physical abuse. At one such meeting, a colleague of Zhou was beaten on the head and lost his eyesight for a month; to help him, Zhou gave him five yuan, a large sum at that time. The two men had to keep the gift secret; otherwise, Zhou himself would be targeted.

Partial relief came when members of the army were sent to manage work-units. While Zhou's life continued to be political study and "confessions" of wrong-doing, the military officer in charge of CCRC allowed the staff to go home at the end of the day; they were no longer required to live in the "cow pen".

Desert and Cold

On November 3, 1969, Zhou and other "bad elements" in the CCRC were ordered to prepare a suitcase and bedcover and board a bus. It took them to Beijing Railway Station. Everyone knew they were being taken to the countryside for "labour and re-education"; but they were not told where they were going nor for how long.

Zhou and his colleagues had a simple dinner at Beijing Railway Station, nobody spoke; their faces were without expression. They boarded the train; they found that all those on board were "bad elements" like Zhou.

The official term used to describe them was "cow devils and snake spirits". In accord with their low status, they were put in third class, on hard seats; that is where they slept. During the journey, no-one spoke. Was there a fixed term for their stay or had they left Beijing and their families for ever? At 63, Zhou was one of the oldest in the group. One of their punishments was that they were not allowed to take books.

After 1,300 kilometres and more than 24 hours on the hard seats, they arrived at their destination and were ordered to disembark from the train. They were in Pingluo county in the Ningxia region of west China. The new arrivals climbed on carts driven by mules, which pulled them slowly through a barren and unpopulated landscape; nobody spoke. After travelling for more than 20 kilometres, they saw a compound with low walls; this was their destination. The walls of the compound were made of mud, with a flat roof; as Zhou discovered, this was because the area had so little rainfall during the year.

This was a "May Seventh Cadre School", one of hundreds of such institutions set up during the Cultural Revolution to "re-educate" hundreds of thousands of intellectuals and officials of the government and Communist Party.

The daily "curriculum" was hard farm labour and constant self-evaluation and study of Mao's works. Mao's idea was to move people from the comfort of their urban life to remote and rural areas; they would work with farmers and "learn the bitterness" of farm life. The name comes from the date of a directive issued by Mao on that day in 1968. They were prisons more than schools; the inmates were severely restricted in what they could do, where they could go and how they could spend their time.

Twenty such "schools" were set up to receive more than 10,000 officials and their families from a dozen departments under the State Council (Cabinet) and the Central Committee of the Party. In addition, provinces, cities, regions and counties set them up to "re-educate" their officials. "We had to swear an oath that we would never return to our original homes," said Zhou. "In fact, it was an oath to say that society did not need people like us. We were the dregs of society. This was the Soviet model. The Soviet Union sent old intellectuals, capitalists and landowners to places north of the Arctic Circle."

Pingluo county could not have been more different to the bustling cities – Changzhou, Shanghai, Beijing, New York and London – where Zhou had spent his life. Pingluo is in the far north of Ningxia, bordering Inner Mongolia to the east and northwest.

Ningxia was a region created by the new government in 1958 for the Hui people; they are Han and Hui by race and Muslim by religion. Most of the region was desert and sparsely populated. It was one of the poorest and most arid places in China. In November, the month Zhou arrived, the landscape around the "school" was desolate – no trees, grass or green shrubs. The winters were bitterly cold, with the temperature falling to as low as minus 30 degrees Celsius.

Most residents of Ningxia were subsistence farmers, living off the sheep they raised and sold for meat, skin and wool. The place had none of the facilities Zhou and his fellow inmates were used to in Beijing – hospitals, pharmacies, restaurants, department stores, bookshops and libraries. How would his health and that of the other elderly inmates survive this desolate place?

They did not know how long they would be detained there, nor what they had to do or say to be allowed back to Beijing. These were some of the many Kafkaesque absurdities of the Cultural Revolution. During the 1950s, the police had sent 10,000 criminals to the Pingluo area to develop farms, as part of the "reform through labour" programme. The place where Zhou and his colleagues were taken was one of the farms that had been developed under this programme.

Good Water, No Leeches

Zhou found that about 70 staff from the CCRC had been sent to Pingluo. He was put in a room with five CCRC men. Inside, he found a stove, made of mud. The six slept close together on top of the stove, known as a kang, which burnt coal throughout the night, to combat the intense cold. The five included a cleaner, a postman and a telegram operator. Each person had brought their own bed-cover. It was Zhou's first experience of sleeping in such a place.

The room was crowded, with little space. His room-mates slept with no clothes on, their custom for many years and a way to keep their clothes neater. In the minutes before they went to sleep, they told jokes and stories about the villages of north China from which they came. For Zhou, the city boy who had lived his life in a world of paper and books, it was a new chapter in his education. His stay in the "school" greatly improved his sleeping habits. Before going to Ningxia, he had slept badly and needed pills to help him sleep. But, in the school, he slept very well – a habit that continued after his return to Beijing. Later during his stay, he was given an individual bed to sleep on.

As he settled in, he discovered that the "school" had some advantages. Pingluo county had three large coal mines; so, coal was cheap and abundant and they did not suffer from cold. Thanks to a hydroelectric power station nearby, they had electric lighting, as well as clean drinking water from a well. Zhou learnt that the school had previously been a "reform-through-labour" camp and that its inmates had been sent to another location, with worse conditions, to make way for the CCRC staff. "Compared to other Cadre Schools, we were better off. We have electric light, good-quality water from a well and could take baths. Because of the weather conditions in Ningxia, we did not have leeches."

His son Zhou Xiaoping and his wife were not so fortunate. Their Cadre School in Hubei did not have such good access to water and electricity and had many blood-fluke snails. Many inmates caught snail fever.

For the first time in his life, Zhou Youguang had to learn how to become a farmer. It was not easy. It meant a five-to-ten kilometre walk over uneven roads to the fields; they often returned to the compound in the dark.

The two main crops they grew were wheat and rice. Since Ningxia was an arid region with little rainfall, most of it was unsuitable for growing rice, which requires large amounts of water. But, because their compound had been established by the central government, they were given a special allocation of water from a nearby reservoir, as well as farm machinery. Such water and machines were not provided to local farmers. As a result, the locals invented a nickname – not the "May Seventh Cadre School" but the "State Council", the name of the Cabinet in Beijing. The local people also laughed at Zhou and his comrades for wearing watches, while they worked on the farm – to them, a watch was an unimaginable luxury.

For Zhou, most difficult was planting the rice seeds, in narrow channels of mud; since shoes would stick, he worked with bare feet. He and the other workers were attacked by hundreds of mosquitoes and other insects, so he had to cover his legs and wear a hat to protect himself.

"After the end of the harvest, because of my age, I was given lighter labour with several lady comrades. We went to the land that could not be cultivated and gathered excreta, from wild donkeys and wild camels. While we were doing this, I could see the wide plains. Only a small part could be cultivated, with wheat and rice. Most could not be developed. This was half-desert with very strong winds and trees could not grow."

They went to collect the excreta early in the morning; it was very cold and they wore several layers of clothes and a scarf to cover the mouth and neck and protect them from the wind and the sand. Once, the wind was so fierce that he had to kneel and close his eyes; the sand beat upon his head like hail stones. When he finally was able to stand, he found that the wind had blown his hat away.

"We Would Never Return to Beijing"

Farm labour was one half of life for "bad elements" like him. The other half was political study sessions, at which everyone had to express an opinion. You could not be silent or say you did not know or understand. The main purpose of May Seventh Cadre Schools was to "transform" the thinking of the inmates into the "correct" Socialist ideology.

One inmate was a lady graduate of Beijing Normal University, unmarried, who had tried to join the Party but had been rejected. She found the conditions of life in the compound unbearable and declared during one session that, come what may, she would not stay there. For this, she became the target of criticism at the next day's meeting. But – unlike Beijing – the sessions did not involve verbal or physical abuse.

"Through these study meetings, everyone came to understand that not only were we 'bad elements' and 'rebels' but also that we were to spend all our lives in the Northwest and that this was our final home. We had to swear that we would never return (to Beijing)."

To improve their understanding of "Socialism", more than 20 inmates went to live and work with local farmers; they stayed with them for a month. But no-one else went after that. Zhou learnt that this was because the farmers were too poor and their conditions of life too difficult – and, probably, the gulf between them and their city guests was too wide. Their lives revolved around the sheep whom they raised. They had no vegetables in their diet. The inmates grew some on the land of the school and gave it to the farmers; they did not know what it was or how to eat it. They had received very little education and were illiterate.

Zhou and the other "bad elements" were considered too "contaminated" to talk to or have direct contact with the farmers. Zhou learnt that, before 1949, the farmers had been wealthier, because the sheep skin they produced fetched a good price. But, after the new government came to power, it took over ownership of the animals and allowed farmers only to raise a fixed number; they became employees of the state, with a salary. They lost their motivation and the number of animals fell. So, they, and the region, became poorer.

Guarding Sorghum

Many sent to the "May Seventh Cadre Schools" were senior officials and, like Zhou, in late middle age. More than 10 in his school had been Ministers or Deputy Ministers. One "classmate" was Lin Handa. He had a distinguished history – a Ph.D. in education from Colorado State University and Vice-Minister of Education in the 1950s; in 1958, he was declared a "rightist" and, during the Cultural Revolution a "counter-revolutionary".

One day, the two men were given an easy assignment – guarding the sorghum harvested by the school. This was because thieves often came and stole the crops. The two were elderly, Zhou 64 and Lin 71. "For guarding the sorghum, there were three rules – you cannot sit down; you cannot stand still but must walk to and fro; and you are not allowed to chat. On the first day, we followed all three rules and did not say a word. But, on the second day, we did not. No-one came to see what we were doing and there was no reason not to sit down. And Lin had much to say; he was full of jokes and stories, which were very interesting."

Lin was not only a distinguished educator but also a linguist and historian who had written best-selling history books. We may ask why the government sent such a talented person to a barren and windswept place in west China to guard sorghum. He died in Beijing in July 1972, aged 72, a short time after the end of his confinement in Ningxia.

Another elderly inmate was Chen Guangyao, a colleague of Zhou's at CCRC, also born in 1906. In the 1920s and 1930s, he had been an early advocate of simplifying Chinese characters; he held important research positions in that department in the CCRC.

"At the school, his health was not good. He often coughed blood, not one or two mouthfuls but a large amount. Even in that condition, he went to work in the fields. Once I saw him spitting blood in the fields. A little later, I saw him back at work." In 1972, a year after returning to Beijing, Chen died. We wonder how much this physical labour contributed to his death.

How did Zhou survive this terrible ordeal? "The Cadre School was very interesting. I turned a bad thing into something enjoyable. I was an optimist and never lost hope. I believed that every bad thing would finally become a good thing."

In the spring of 1971, the inmates were allowed to return to Beijing for 10 days to see their families. When Zhou went to his son's house in Zhongguancun, he found that he was not there. He and his family (including Zhou's wife) had moved into the large living room of a four-room courtyard house of one of his teachers. The teacher invited him to move, in part because he liked Zhou's son and in part because, if the room was not filled, others would move in and it would be hard to get them out. So, Zhou spent his 10 days in Beijing in this living room.

In an interview in 2017 after Zhou's death, his granddaughter Zhou Heqing described his state of mind during his visit; she was 12 years old. "He was warm and friendly. Not allowed to take books to Cadre School, he took 20-30 copies of 'Mao Zedong Works" from different countries and in different languages which he could compare; he also took a *Xinhua Dictionary* to research the characters. I respect Grandfather. Despite the Cultural Revolution and the Cadre School, he did not give up hope. He was optimistic and hard working, as before." It was this spirit of forbearance and intellectual curiosity that enabled him to survive his privations and live to the age of 111.

When Zhou returned to Pingluo county, he found the atmosphere greatly changed. Almost no-one was going out to work, and nobody asked him to. Elderly inmates like him were no longer sent to the fields. Instead, they were given lighter tasks indoors, like cleaning tools.

Wisely, Zhou had brought from Beijing a large Chinese dictionary. He used his new free time to analyse the composition of the characters; there was nothing "counter-revolutionary" about this, so no-one interfered with him. After he returned to Beijing, he would turn the material into a book. As things became more relaxed, more and more people came to talk to these "bad elements" and in a more polite way. Men and women began to flirt and tell jokes to each other. Even the local farmers came to talk to Zhou. They asked him about his rich experience of living abroad and for his advice as an economist on how to develop Ningxia.

One spring day, he and his roommates were summoned at five o'clock in the morning to assemble in a large square in the centre of the school for a meeting. While it was cool at that hour, Zhou knew that, if the meeting lasted several hours, the sun would become hot; so, he wisely took a wide-brimmed straw hat to protect himself from the heat. In the event, it was not the heat that threatened them. The school was close to a habitat for wild geese; they liked remote places, not those full of humans, like big cities.

"At 10 o'clock, the sky was filled with a huge flock of wild geese, not several thousand, but tens of thousands. They flew to and fro above us inmates in the square. They were highly organised, in formation like a squadron of jet fighters, with a leader in front. The sky became dark as they flew overhead. On the loud screech of a leading goose comrade which no-one had heard before, they all excreted at the same time on the assembled crowd; it fell down on us like a rainstorm. Since I had the hat, the impact on me was limited, but not my comrades, who were covered from head to toe in goose faeces; it was very hard to remove. One local resident said that, while such behaviour by the birds was normal, such a large discharge on the heads of people happened once every 10,000 years. This was a strange experience, something which I have never forgotten."

Then, one cold day in mid-September 1971, Zhou and his colleagues were summoned to an emergency meeting. They heard a report from Beijing so extraordinary that they found it almost impossible to believe. Lin Biao, Defence Minister and successor of Chairman Mao, had died in an aeroplane crash over Mongolia on September 13, together with his wife and son; they had been planning an armed rebellion against the government. Lin had commandeered a plane in Qinhuangdao in east China; it had run out of fuel and found nowhere to land in the middle of the Mongolian steppe.

On hearing this news, everyone in the room put their hands up to ask the same questions – "How could Lin Biao be a traitor? Why would he plan a coup?" Zhou's comment: "At that time, no-one knew it was true or not. According to the Constitution, Lin was the person appointed as the successor (to Mao)."

Few held a higher place in the Communist pantheon than Lin Biao. As a general, he commanded the People's Liberation Army that defeated the Kuomintang in Manchuria from 1946-48 and led his troops into Beijing. He ranked third among the Ten Marshals of the People's Liberation Army. He had been Minister of National Defence from 1959 until his death. In 1966, Mao named him as the single Vice Chairman of the Communist Party, making him his successor. He had worked closely with Mao since 1928, when they were part of the same soviet in Jiangxi province, southeast China.

The incredulity of Zhou, and millions of other Chinese, was understandable. Since the start of the Cultural Revolution in August 1966, nobody had praised Mao Zedong as warmly as Lin Biao. A narrative later released by the government said that, from February 1971, Lin and his wife began to plan the assassination of Mao. In early September, Mao was returning to Beijing by train and Lin ordered an attack on the train; but Mao changed his route and the attack failed. In panic, Lin Biao and his family attempted to escape to the Soviet Union and crashed en route near Ondorkhaan in Mongolia.

In February 1972, Zhou heard another piece of startling news. U.S. President Richard Nixon had arrived in Beijing, to a warm welcome from Chairman Mao. The process of normalisation with China's bitterest enemy since 1949 had begun.

These two world-shaking events brought a great benefit to Zhou and his fellow inmates – they were able to return to Beijing. In April 1972, the Cadre School closed. The inmates turned the institution over to the local residents, who were delighted. They called it "the State Council" because it had conditions far better than their own.

It is hard to know the links, if any, between the death of Lin Biao, the arrival of Richard Nixon in Beijing and the closing of the May Seventh Cadre School in Ningxia and elsewhere. During the Cultural Revolution, Lin had been the most prominent supporter of Mao. His betrayal and flight made the talk of "Socialist" brotherhood between them meaningless. Had the May Seventh Cadre Schools in Ningxia and all across China served a useful purpose? Had the inmates been "re-educated"? What was the cost to their mental and physical health? What was the loss to China's government and academic community of having so many talented people detained in the countryside? It was best not to pursue such questions.

After staying with his wife in Suzhou, Zhou returned finally to Beijing in the spring of 1972. He had spent a total of 28 months in Ningxia.

In an interview in Beijing in September 2013, Zhou said that, if the death of Lin Biao had not happened, he and his fellow inmates would not have been able to return to Beijing.

"After coming back from the Cadre School (in Ningxia), we high-level intellectuals were summoned to the State Council and told: 'You people are the dregs of society and are useless. For humanitarian reasons, we give you food to eat. Go home and do not speak or act improperly. We had no work and returned home. Communism denies intellectuals and they must be eliminated. During that period of history, I had no value."

CHAPTER NINE

Return To "Normal", Pinyin Becomes Global Standard

In the early spring of 1972, Zhou and his wife finally returned to Beijing. He had been away for two years and four months. But they could not live in their home in the compound of the CCRC. It had been taken over by others; this was common during the Cultural Revolution. Red Guards and others occupied the homes of people designated as "black elements", saying that they had lost their right to live there.

With the breakdown of government across China, it was difficult to seek redress, especially for those like Zhou expelled from Beijing for a long period. So, he and his wife lived for a long time in the house of their son's teacher; earlier, the son's teacher had invited the son, wife and child to stay there to prevent others from moving in.

Zhou approached his superiors at CCRC to get his home back. After negotiating with those who had moved in, the superiors were able to obtain two of the four rooms in the home for Zhou and his wife. These were two small and medium-size rooms; the two larger ones remained with the new occupants. "But we were already very satisfied," Zhou said. How well he adapted to the new "normality".

They found that the Red Guards had stripped their home of everything – books, articles, notes and photographs. Not one piece of paper remained. For a period, his wife lived with their granddaughter in the home of a friend, the Vice Minister of Commerce; she did housework and worked as a maid.

It was the second time they had lost everything. In early 1946, when they returned to their home in Suzhou, they found that everything in it had been removed and strangers had moved in. There was no trace of the man to whom they had entrusted the house in 1937.

"During the Sino-Japanese war, there was death and escape. Compared to the suffering of that war, the Cultural Revolution was a minor matter. I had no attachment to the items in our house. This attitude was a big help. Our feeling toward assets was very shallow. We felt that it was something outside the body. There is a saying in Buddhism: 'If you regard items outside the body as very important, your spirit will suffer pain.'"

Zhou showed remarkable calmness toward the loss of his books, papers, documents and records. The thing he found hardest to accept was the loss of the family photographs – they could never be replaced.

"My impression after returning to Beijing was that the central and Beijing governments were paralysed. The government was re-organising itself but seemed not to be in a hurry. It was as if the public had just recovered from a grave illness and was not so healthy. Everyone was unclear about the future," he wrote.

Zhou returned to the CCRC but was given no work to do. He attended study sessions and meetings. Those in charge were the same people who had taken control during the Cultural Revolution. Zhou continued his own research at home; fortunately, while he had lost most of his books, he still had some on linguistics and Chinese characters.

Roving gangs of Red Guards no longer came to disrupt him; he could work quietly. The walls and streets of Beijing were kept clean, without big-character posters. He also taught his granddaughter English and how to type; he bought her many sets of English learning books. His wife taught her Classical Chinese. She also learnt the violin for several years. Still a "bad element", Zhou was very careful what he said and did in public, to avoid becoming a target for struggle and criticism.

The upside was that, since he had no work at the CCRC to do, he could concentrate on research – which led to several books being published after the Cultural Revolution; without this absence of work, he would not have been so productive. He published three books on linguistics in 1980.

Zhou's income was 35 yuan a month, not enough to live on; he had to borrow to support himself, his wife and daughter. His debts reached 4,000 yuan. Then, one day, his bosses informed him that, since he had returned from the Cadre School, his salary would return to the pre-1966 level and that they would pay him the money owed to him for the last six years. Suddenly, he felt rich, and did not need to worry about money. "The political and social atmosphere was as if everyone was waiting for something to happen," he said.

Success of Pinyin

In 1972, the CCRC had been abolished as an institution and replaced by a small committee. For three years, this committee worked on a revision of the Pinyin system which Zhou had created. The committee criticised it for being too complicated and was devising a simpler system. The committee did not ask Zhou to join but constantly asked his opinion.

He adamantly opposed their revision, saying that Pinyin was not complicated. He told them that, since the National People's Congress had approved his system, it would also need to give its approval to a revised system. Faced with opposition from many quarters, the committee failed in its attempt. Zhou was delighted.

Another example of his success came in revisions of a new Xinhua Chinese dictionary and an official textbook for primary school students. The issue was whether to add Pinyin above the characters. There was intense debate, with strong arguments on both sides; the matter had to go up to Prime Minister Zhou Enlai himself to decide. He ruled in favour of adding Pinyin. "These two things were evidence that Pinyin was something that was essential and could not be easily changed. This was the success of Pinyin," Zhou wrote.

Earthquake and Protests in Tiananmen Square

In April 1976, Zhou found that he had a problem with his prostate that required medical treatment. He went to the Capital Hospital, one of the most famous in the city. It was founded in 1921 by the Rockefeller Foundation as the Peking Union Medical College Hospital. For several years during the Cultural Revolution, it was called the "Anti-Imperialist Hospital". Then it reverted to Capital Hospital. The doctors decided that Zhou needed a period of observation, followed by a major operation.

Ever industrious, while he was in his hospital bed, Zhou worked on a book on Pinyin, which was published in 1980. In the next bed was a man whose wife worked in a Beijing factory. Every afternoon, she came to visit him; her route from the factory took her through Tiananmen Square, and she described the dramatic events taking place there. On January 8 that year, Premier Zhou Enlai had died; he was the most popular national leader. Ching Ming Festival, which fell on April 5, was the day when Chinese mourn their dead. As this day approached, the wife reported, many people went to the square and left wreaths of white flowers around the Monument to the People's Heroes in the square. Zhou was not allowed to go and see for himself.

The wife explained that this mourning was an implicit criticism of the Gang of Four, the group of Communist Party radicals who implemented the harsh policies of chairman Mao Zedong during the Cultural

Revolution from 1966 to 1976. On April 5, she described that she had seen police dispersing thousands of people mourning Zhou and implicitly criticising the government. During the night, the police came and removed all the wreaths. It was a repeat of what had happened on May 4, 1919, when thousands of students protested the terms dictated to China at the Versailles Peace Treaty, and what would happen on the days leading up to June 4, 1989. "Whenever I look at the Pinyin book (he wrote at that time), I think of April 5 in the square," he wrote.

Zhou remained in hospital for three more months; the condition of his prostate worsened. His doctors decided to remove it. In the early morning of July 28, he and the other patients in the ward were awoken by a fearful shaking of the building. He went to the window and saw the sky covered in red. He heard people outside shouting "Earthquake, Earthquake". These were after-shocks from a terrible quake which destroyed Tangshan, an industrial city east of Beijing. Zhou told the others in the ward, which included his wife and granddaughter, to take refuge outside, in case the entire building collapsed. "We were very fortunate. Some in the hospital were injured, but nobody died."

His family members also survived unscathed. Elsewhere in Beijing, many died from collapsed trees and buildings. The next day came another trial – heavy downpours of rain. Fearful to return inside the hospital, Zhou and his fellow patients stayed outside; they erected a tent on the back of a lorry and took shelter inside it. The hospital managers advised him to leave the earthquake zone and go to the south, so they purchased seats on a flight to Shanghai.

They went to the airport – but found that their plane had not arrived. In those days, air travel in China was limited. As they waited, a plane arrived from Tangshan; its people were seeking the safety of Beijing. Zhou and his wife heard first hand from the arrivals the terrifying experience they had lived through – blocks of buildings collapsing like playing cards, dead bodies so numerous that they could not be counted and nowhere to bury them. Some were put in plastic bags, taken on an aeroplane to Tianjin and dropped into the sea. Of the arrivals, some were weeping, some trembling with terror and others eager to describe their ordeal. It was the deadliest earthquake in the history of China, registering 7.8 on the Richter scale. Within minutes, 85 per cent of the buildings in Tangshan collapsed, including rail and road bridges. The official toll was 242,769 dead and 164,851 seriously injured.

When Zhou reached Shanghai, he found that some people there were also seeking to run away, even though tremors there had been minor. The city was full of earthquake rumours. Zhou and his wife stayed in the home of a family member. "In the middle of the night, a friend made a call to me to announce a big event – he said that Mao was dead. There was no television in those days. We could only listen to the radio and learn that this was true. The radio broadcast mourning music." Mao died 10 minutes after midnight on September 9, 1976. The government did not announce the death until 16:00 that day.

On October 6 came more important news – arrest of the "Gang of Four", the main leaders of the Cultural Revolution. "The home of my relatives did not have a telephone. So, I went to the main entrance and called several close friends in Shanghai. That evening I was able to confirm the news."

The news electrified the city. The next day it was full of big-character posters denouncing the evils of the Cultural Revolution; anger bottled up for 10 years suddenly burst out into the open. But state radio and television did not report the posters. Like many Chinese, the broadcasters were uncertain and fearful what direction the government would take; it was safer to stay silent.

Because Zhou had not fully recovered from the removal of his prostate, he decided to spend the winter of 1976/77 in Suzhou; it was quieter and more peaceful than Shanghai. The house where the family had lived before 1937 had been destroyed during the war. Before 1949, his wife's family had owned many properties in the city; but these had been taken over by the new government. Fortunately, a close friend of his wife, a primary school teacher, took in the Zhous and lent them rooms in her house. To help them through the winter, friends provided a stove. But where to buy coal?

Like the rest of China, Suzhou was part of the planned economy. Goods were rationed; to buy them, you needed coupons as well as money. Suzhou had 51 different coupons. Fortunately, the sister of Zhou's landlady, a Suzhou resident, received the coupons; but she was away from the city that winter. So, the Zhous were able to use her coupons to buy coal.

In Beijing, they had been used to shortages, even though supplies were more plentiful in the capital than elsewhere. In Beijing, they had to get up at 05:00 and join a long queue stretching onto the street to buy vegetables. But, by the time they had bought them, the meat was sold out. So, friends bought more meat than they needed and traded it for vegetables the Zhous had bought.

In Suzhou, as elsewhere, all retail trade was state-owned; private commerce was illegal. People made their own arrangements. One day a lady farmer arrived at the door of the Zhou house with a basket of vegetables. Once the door was firmly shut and prying eyes could not see inside, the lady removed the contents and the family was delighted to buy them. One day Zhou saw someone selling roasted potatoes on the streets; he was happy to buy some. But police soon came to arrest the hawker – this was "capitalism".

One pleasure of Suzhou was to eat hairtail, the fish Zhou had most enjoyed during his childhood. They lived only on a section of the Yangtze River near Suzhou and were unavailable in the city's fish markets. The only way to purchase them was to get up before daylight and buy directly from the fishermen. Amid the chaos around him, eating hairtail was one of Zhou's happiest memories of this tumultuous period.

As the months passed, Zhou's health recovered. He went to Dongting Dongshan, next to the Tai Hu lake, one of the favourite holiday spots of his childhood. Its fishermen caught crabs that are the best in China. They told Zhou that they had sent the crabs directly to Beijing, for the dining table of Chairman Mao. Zhou met the chief of the local commune; this had been the basic unit of agriculture in China for more than 20 years. He told Zhou his farmers lacked motivation and that their superiors gave them unrealistic targets. As Zhou was a member of the Chinese People's Political Consultative Conference (CPPCC), could he mention this to the central government when he returned to Beijing? He agreed.

The next morning, before sunrise, Zhou went to see the farmers preparing to go to the fields; they were listless and walked slowly – no-one wanted to work. However much, or little, they worked, they received the same pay. Their chief was right.

For Zhou, the root of the problem was collectivisation of agriculture and the system of People's Commune. "If the People's Communes were not abolished, then China would not have had food to eat." He said that collectivisation was a very primitive form of agriculture popular in the

Spring and Autumn Period (770-476 BC), in the interim between the slave and feudal systems, with very low productivity. In this system, the peasant paid his rent in the form of labour.

During the early stage of rural collectivisation in the 1950s, Soviet experts came to Beijing and told the new government that this experiment had failed in the Soviet Union; they advised the government not to do it. "At that time, I did not fully understand what they meant," wrote Zhou. "But seeing the situation in the Suzhou People's Commune left a deep impression on me."

No Longer a "Black Element"

His health recovered, Zhou returned to Beijing, back to his own house: small, but with enough space for him to continue his research in peace. During the Cultural Revolution, Kunqu Opera had been condemned. But, from the autumn of 1978, his wife and other enthusiasts were able to revive it. Officially, the Great Proletarian Cultural Revolution had ended in the winter of 1976, after the death of Mao and the arrest of the Gang of Four. But, Zhou said, this was not correct. He said that it took until 1987 before things fully returned to normal. "All this was very slow." Officials installed during the CR kept their posts and their "revolutionary" ideas.

One example was the CCRC, whose leaders did not change after the Cultural Revolution. They continued to pursue a project known as "The Second Simplification" of Chinese characters. As we described in Chapter Six, the government had in 1958 approved a sweeping reform of the characters, to make them simpler and easier to understand. It was the second major reform of the CCRC, at the same time as the introduction of Zhou's Pinyin system. The revolutionaries in charge of the CCRC believed the first simplification did not go far enough; they drew up a proposal outlining further sweeping simplifications. They did not consult Zhou and other language specialists because they were "capitalists"; this disqualified them from the work.

Aware that their power was ebbing, the CCRC leaders rapidly submitted their proposal to the State Council (Cabinet). It provoked a flood of opposition from within China and abroad. Many wrote their opinions to Zhou. Now that he was no longer a "bad element", he was able to publish his views in the newspapers. So, he wrote an article for the *Guangming Daily*, the main newspaper for intellectuals, outlining his opposition. He said that a second major reform coming just 20 years after the first one was too fast; it would need meticulous research by experts. The public had become accustomed to the characters set out in the first reform and did not want more changes. Thirdly, many of the changes made the characters more confusing, with two or more meanings; some of the new words introduced were used only in one part of China or in one profession and not others.

Such was the torrent of opposition that the proposal was finally abandoned in 1986. It was a victory for Zhou and other specialists in his field. That year the CCRC was renamed the National Language and Character Work Committee. China has maintained until today the system of Pinyin and simplified characters that Zhou and his colleagues at the CCRC devised in the 1950s.

After the Cultural Revolution, one of Zhou's jobs at the CCRC was to receive foreign guests. At that time, not many people in China had his foreign language and social skills. His granddaughter said that these guests gave him gifts; he did not bring them home but left them in the official car for the driver to give to his family. She said that his colleagues would steal cinema tickets and other items given to him. "He would say:

'Do not waste time on these things. Let us save our time and energy for study," she said.

Looking back on the Cultural Revolution, Zhou had a strong opinion. The official verdict is that it was "a ten-year catastrophe" that lasted until the end of 1976. But, Zhou said, it continued for several years longer, lasting to the early 1980s and no fundamental change until 1985. "So, it actually lasted for 15 years. What was the purpose of this catastrophe? I do not understand." He put this question to many people in high positions. Some replied that Mao did not trust at all the political system created by the government after 1949.

At a meeting in Shanghai in 1966, Mao called for a continuation of "revolution"; all the provincial and military leaders opposed him, except Lin Biao. For this, Mao rewarded him for his support by naming him as his successor.

One of the thousands of deaths during the Cultural Revolution was that of Liu Shaoqi, Chairman of State between 1959 and 1968. He died on the morning of November 12, 1969, while he was living under an assumed name in Kaifeng, Henan province. "Liu's death was very tragic. It is said that, when he died, he was naked. His body was wrapped in a mat and thrown into a crematorium without a name ... I do not understand. Why was there destruction on such a large scale? The whole education system was shut down for 10 years, influencing an entire generation of people. The Cultural Revolution caused a loss of trust in the Communist Party. This crisis of trust was very frightening. In the early 1950s, the country's political system was stable and the government was methodical. Under the leadership of Zhou Enlai, people had trust in the government. But one movement after another destroyed this trust – including the anti-rightist movement and the People's Communes."

The concentration of power in the hands of one man, Chairman Mao, was like that of many emperors in Chinese history. "So, some people said that Chairman Mao achieved two great things in his life. One was to create New China. The other was to destroy it."

In March 1978, Zhou resumed his place in the CPPCC, 13 years after his last meeting. He became the deputy chairman of its education group.

Pinyin Becomes Global Standard

One cold morning in early 1979, Zhou was surprised to receive a visit from his superiors; they told him that the next week he would go to Paris to represent China at a meeting of the International Organization for Standardization (ISO). He was surprised because he thought he had not been fully "rehabilitated"; he was able to work and receive his full wages but had not been fully accepted in the CCRC. Such a person should not be able to go abroad in an official capacity.

"I told them that I did not wish to go because I had no decent clothes. They said that would not be a problem. I should buy a new set, the best quality. But, when I returned, I must give the clothes back to the government. They sent two people to escort me to the airport. There they told me that it was an invitation from the United Nations, which would give me a lot of money, and so they would not give me any U.S. dollars. This money I must give to the government. They also told me to empty my wallet because I was not allowed to take renminbi out of the country. Our government was absurd to this level. The Communist Party did so many absurd things. I felt that it was very amusing."

The journey would be his first visit outside the mainland since 1949. During those 30 years, he had not exchanged a single letter with anyone abroad; it was too dangerous to have such contacts.

Zhou's was an important mission. Founded in February 1947, the ISO is an international organisation in Geneva that develops and publishes standardisation in technical and non-technical fields. As of 2022, it had 167 member countries and had developed over 24,000 standards, from manufactured products to food safety, agriculture and healthcare. Zhou's task was to make Hanyu Pinyin the international standard for written Chinese. This had great significance, to make the system he had invented the global standard, to be used in personal and place names, scientific products, publishing of books and magazines and elsewhere. Before 1979, Taiwan had represented China at ISO meetings; so, Zhou was going as the first representative of the mainland government.

Zhou received from UNESCO a return Beijing-Paris ticket on Air France. Because of the official policy, he boarded the plane with his pockets empty – no renminbi or U.S. dollars – and uncertain what would happen at the other end. Under such circumstances, another person might have refused to go – no money to hire a taxi or even make a telephone call in Paris. But Zhou had lived through enough in his life to take this uncertainty in his stride.

At Paris airport, he was, fortunately, met by the wife – and secretary – of the Chinese ambassador and her chauffeur. They drove him to the embassy guest house, where he dropped off his bags, and to the office of UNESCO to collect his expenses for the visit. There, after giving the money, the official asked him: "We have invited you for three years. Why did you not come?" Zhou did not know how to answer; he had received no invitation letter, probably because his superiors withheld it from him. So, he kept silent.

One Chinese was enough for the ISO meeting, but the government had sent three more with him, plus a translator of English. The problem was that, unlike Zhou, none had a mastery both of English and the technical subjects under discussion. This was not surprising. For 30 years, China had discouraged – and often banned – contact between its citizens and the outside world; the only people allowed to learn foreign languages were students at specialised institutes trained for government departments.

With the open-door and reform policies, things gradually improved. These were policies implemented by Deng Xiaoping from the early 1980s that gradually opened China to the outside world and permitted the private economy. In 1984, the government issued a new rule – those attending international meetings of a technical nature had to understand both the subject being discussed and a foreign language.

The Paris meetings were a preparation for the main session to be held in the Polish capital of Warsaw. While English was the sole language used in Paris, the meetings in Warsaw were conducted in both English and French. Zhou prepared two articles for the conference. One was on the historical background of Pinyin and the other on its technical features, especially why they had created a new form of Romanisation to replace the Wade-Giles system used for nearly 80 years.

He gave two presentations on these subjects to those attending and was applauded. One delegate told him that he had taken part in many meetings on the issue of how to write Chinese but none had produced a result; no-one had presented the subject so clearly as he. Zhou condensed the articles into one; it was published by a technical magazine of UNESCO, in its third issue of 1979. It paid him an allowance of US$300 for his time and work. For a Chinese at that time, this was an enormous sum of money.

Moscow – Faulty Lifts and Swinging Doors

After the Warsaw meeting, he went back to Paris. There he gave the money to the wife of the ambassador, saying that it was not easy to take it back to China and he had no use for the money there. She could buy books for him. Two days later, she told him that, in accordance with official regulations, she had handed the money over to their government, and so could not buy books for him.

But how could he return home from Paris without a gift for his wife? He was allowed to spend the foreign exchange equivalent of 20 yuan to buy something to take back. His wife had asked for a green floral bed cover. The ambassador's wife drove him to a Paris discount shop; there he found a green flowered nylon bed cover. "It was very beautiful and we used it for many years."

In Paris, he had limited time for tourism. But he was able to meet his brother-in-law, Zhang Ninghe. He had gone to study music in Paris; there he met a Belgian pianist who became his wife. After graduation, they went to Beijing where he worked as a conductor with the Central Orchestra.

As China moved to the left, Zhang had to leave this elevated position because he had a foreign wife. He moved to a company that made music for films. He and his wife made regular visits to Belgium to see her parents.

After the outbreak of the Cultural Revolution in 1966, he applied to the police for permission to visit Belgium; they refused, saying that a world war would break out soon. So, he wrote to Premier Zhou Enlai, whom he knew, asking for permission; Zhou granted it.

After this experience, Zhang and his wife did not return to China; they feared that, if they did, they would not be allowed to leave again. During one of the meetings in Paris, Zhang brought his son, a primary school student. Since it was not holiday time, the son had homework to do; he brought a typewriter and wrote his composition in the car his father was driving! Zhou enjoyed the sights of Paris, including the Centre Pompidou.

To take best advantage of this rare visit outside China, Zhou changed his air ticket after returning to Paris. He traded in the Air France ticket for a Lufthansa flight to Moscow and, from there, an Aeroflot plane to Beijing. This gave him an opportunity to see Moscow, a city he had never visited; it was the birthplace of the Communist revolution.

Having spoken at a UNESCO conference on behalf of China, he was a person of stature. The embassy in Moscow put him up in its guesthouse; it was large, but the number of visitors was small. The embassy was an imposing structure on Druzhby Street in the Ramenki district. Both the embassy and guesthouse were the children of the 10-year honeymoon between the Soviet Union and China after the 1949 revolution. Ideological differences drove the two countries apart after 1961. So, when Zhou arrived, the mood was tense.

To his surprise, the ambassador asked him to give his staff a talk about what he had seen in France and Poland and events in China. Zhou had imagined that an embassy in such an important country should be extremely well informed. "We are not free in our movements," the ambassador said. "What we know about China and the international situation is very limited."

That evening nearly the entire staff of the embassy attended the meeting where Zhou spoke. As soon as he started to chat to the diplomats, they shut the door, fearful that someone outside was listening. "I was not a specialist in politics. I knew only a little about China and the outside world from what I read in domestic and foreign newspapers in China. I told them everything I knew and offered my opinions. They were very happy."

He was impressed by the size and grandeur of buildings in Moscow but also the poverty. There was a repair man for each lift of the guesthouse he stayed in; one of them told him that, when the lift had a fault, the passenger rang a red bell and he repaired it at once. The rooms were very large but the door hinges were not straight. On his final day, he had a few rubles left over and went to Moscow's largest department store. But, when he asked to buy something, the staff said that it was only a sample and they had no stock. Unable to spend the money there, he spent it on trinkets outside.

The Aeroflot flight to Beijing was also instructive. To save energy, all the lights in the passenger and cargo sections of the large aircraft were turned off. There was only a handful of passengers; each put down the seat rests and was able to have a good sleep lying down. "If this was how the airline ran, it certainly lost money and needed a subsidy from the government."

The People's Liberation Army Guesthouse

The warm reception to Zhou and Pinyin at the Warsaw meeting led to a request that the ISO formally adopt it as the global standard for written Chinese. China was asked to host a meeting of the ISO. In 1980, the government agreed to such a meeting; but it did not want it to be held in Beijing, because it feared too many "spies" among the foreign delegates. Instead, it chose Nanjing and selected as the venue a guest house of the People's Liberation Army on a remote and quiet hillside far from the city.

It was among the first such international academic meetings in China after 1949. There were many headaches, including a lack of copying machines. Several were specially flown to Nanjing from Beijing, complete with two repair men in case they broke down.

The meeting passed a resolution to accept Pinyin as the global standard, with a document explaining what it was and the reasons for accepting it. The directors of TC 46 under ISO approved it and sent it to member countries to approve.

The representative from Taiwan did not attend the Nanjing meeting, and his government voted against the proposal; it had its own Wade Giles system for writing Mandarin. Following its lead, both the United States and Britain voted against. But more than three quarters of the member countries voted in favour, so the proposal passed.

In 1982, Pinyin became the international standard, with the number ISO-7098. It was a milestone in Zhou's life and work. It was recognition not only from his own government but also the international community. In 1986, the United Nations adopted Pinyin. In the early 1980s, Singapore adopted Pinyin for teaching Chinese in its schools, shortly after launching its Speak Mandarin Campaign in 1979. As the influence of the People's Republic grew around the world, so did Pinyin.

In the U.S., agencies of the Federal Government, the scholarly community and the media adopted Pinyin. On October 1, 2000, the U.S. Library of Congress started to use Pinyin. In a statement, it said: "The Wade-Giles romanization system, followed in American libraries for the last century, will no longer be used. In Pinyin, the former Chinese leader is called 'Mao Zedong,' as opposed to 'Mao Tse-tung' under Wade-Giles; 'Qing dynasty' (Pinyin) will be used, rather than 'Ch'ing.' Now, the Library of Congress and other U.S. libraries are synchronised with the romanization used by other U.S. government agencies, including the Board of Geographic Names – the body that governs the form of geographic names used in Library of Congress cataloguing. In making the change to Pinyin, the Library collaborated with Online Computer Library Center (OCLC) and the Research Libraries Group (RLG) to address the conversion of the millions of Chinese-language bibliographic records in their respective databases romanized according to Wade-Giles, including the headings established from Chinese works that exist in non-Chinese records (e.g., translations of works by Mao Zedong). 'The adoption of a single romanization scheme will enhance the flow of information around the world, helping both librarians and library users wherever they are,' said Jay Jordan, OCLC president and chief executive officer."

It was no simple task – the move to Pinyin took two years of planning and co-ordination. The Library of Congress is one of the largest libraries in the world, with more than 119 million items in more than 460 languages. Pinyin was also accepted by the American Library Association and many other international institutions. In Europe, governments, institutions and universities also adopted Pinyin.

Pinyin has played an important role in facilitating China's entry into the family of nations. A system of writing Chinese in Roman letters was essential. For example, if a foreigner was seeking a Chinese book or magazine, he simply typed the name and the author in Pinyin and was able to find it quickly. Pinyin became the main tool for foreigners – as for mainland primary students – to learn Mandarin. As of 2022, an estimated 30 million people around the world were studying the language; the vast majority using Pinyin. Without it, it would be very difficult for them. Pinyin has also played a critical role in bringing Chinese into the era of computers and the Internet, as we will discuss in Chapter Eleven.

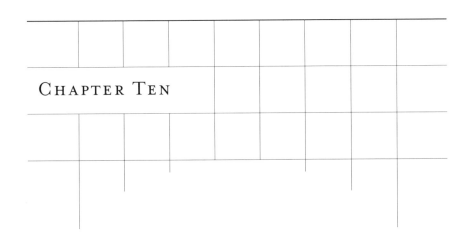

CHAPTER TEN

"Encyclopaedia Zhou" Hong Kong And The United States – Reconnecting With The World

Between 1980 and 1985, Zhou left China several times, to visit Hong Kong, Hawaii and the United States. Between October 1984 and February 1985, he spent four months in the U.S., travelling from California and the Mid-West to New York. Famous as the scholar who had created Pinyin, he was in great demand as a lecturer; he spoke at universities in Santa Barbara and Yale and at the United Nations. It was an opportunity to meet scholars of Mandarin, Chinese and non-Chinese, including those from Taiwan, from which the mainland had been cut off since 1949.

During this period, his main project was as one of three editors charged with translating 10 volumes of *Encyclopaedia Britannica* into Chinese. His network of friends and contacts in the United States was remarkable, despite the fact that he had been unable to leave China for 31 years after 1949 and communication with the outside world was difficult, if not illegal, during much of that period.

Hong Kong Replaces Shanghai

In October 1980, Zhou made his first visit to Hong Kong in 31 years. In 1949, it had been his final stop on his return from Britain to Shanghai. He was invited to speak at the China Language Society that had been set up in Hong Kong after the Cultural Revolution. The invitation came from a mathematics professor at Hong Kong University and the son of a director of the Xinhua Bank where Zhou used to work; the son was familiar with Zhou's scholarship and creation of Pinyin.

The policies of Mao Zedong had created a wall between the mainland and Hong Kong; they were almost like separate countries. During the political movements, especially the Cultural Revolution, people in Hong Kong did not dare to visit the mainland, for fear that they would not be allowed to leave. Instead of looking to their mother country, they turned the other way, to Britain, Europe, the United States, Canada and East and Southeast Asia. They were the markets for goods manufactured by Hong Kong's factories and where its people went to study, take holidays and emigrate. This was reflected in the languages used in Hong Kong. The official language of the government, the law, finance and much of business was English; the vast majority of the local population spoke Cantonese. Mandarin, the national language of both the PRC and Taiwan, was restricted to intellectuals and scholars, a small number of schools and those working for mainland companies. When Hong Kong people saw thousands of Red Guards chanting Maoist slogans in Mandarin, they were repelled.

Zhou was astonished by what he saw in Hong Kong. In 1949, it had been a backwater compared to Shanghai, then the financial and industrial capital of China. Fearful of a Communist takeover, many British people were leaving. But, he discovered, the roles had now been reversed. It was Hong Kong that had become a major financial centre and city of wealth; Shanghai had stagnated.

Zhou and his wife stayed in the Young Men's Christian Association (YMCA); he found it extremely comfortable. There were few hotels of that standard in Beijing – but, in Hong Kong, it ranked only in the middle range. He gave a lecture at the University of Hong Kong. He found it modern and advanced, with funding far greater than in universities in Beijing. "My visit to Hong Kong left me with a deep impression over its economy. Over the decades, there had been an enormous change. It was truly a historic transformation."

He found that a majority of the Hong Kong public were patriotic to China but did not like the Communist Party; loving China did not mean loving the party. The majority of Hong Kong residents were refugees from the mainland who had escaped Communism, and their descendants.

Zhou was very well received. He gave lectures on the reform of the written language, the simplification of the characters, Pinyin and promoting Mandarin. His listeners taped and recorded them, and he was introduced to many different people. The language society which invited him put on the front page of its magazine a list of the more than 200 articles he had written since 1949. All this showed that Hong Kong people were eager to learn the development of Mandarin during the previous 30 years.

He was also delighted to find that the Xinhua Bank, for which he used to work, had expanded to 77 offices across the city, compared to two in 1949. Its general manager invited Zhou to a dinner at an expensive rooftop restaurant, where an elegant lady photographer took pictures of him.

Pinyin Conquers the World

Pinyin made great strides across the world after the ISO adopted it as the global standard for written Chinese in 1982. The United Nations adopted it in 1986. It was accepted by the U.S. Library of Congress, the American Library Association and many other international institutions. In 2000, the Library of Congress spent US$20 million to convert to Pinyin the romanisation of its 700,000 Chinese volumes from the Wade-Giles system it had used since 1957.

Hawaii

In 1983, Zhou returned to the United States for the first time since 1946. He was invited to take part in a major event at the East-West Center in Hawaii in September, "The International Meeting on the Modernisation of Chinese".

It was a landmark event, made possible by Beijing's open-door and reform policies and an improvement of relations between the mainland and Taiwan. The organisers invited more than 60 scholars of Chinese from 11 countries. They included nine from Taiwan, of whom five took part; and six from the mainland. One was refused permission by his government to go, so the other five took his thesis. It was the biggest international event of its kind since 1949.

There was much to discuss. Chinese was the most widely spoken language in the world and had since February 1946 been one of the five working languages of the United Nations (a sixth, Arabic, was added in December 1973). But, unlike the other five, Chinese did not have a standard form. The mainland and Singapore used simplified characters and Pinyin. Hong Kong and Taiwan used traditional characters; Taiwan had its own phonetic system, Bopomofo. Hong Kong, Taiwan and the mainland often used different characters for the same foreign name, such as U.S. President George Bush. By contrast, L'Academie Francaise had been founded in 1635 to standardise spelling, pronunciation and meanings of the French language; its rulings have been final for the use of the language across the world. There was no comparable institution for Mandarin. This was the legacy of 60 years of political and linguistic conflict in the Chinese world.

Getting to Hawaii was no simple matter. At that time, foreign exchange in China was strictly controlled. Zhou applied for the money to cover the return air fare; but it was not approved because he and the other four were not being sent by the government. As the opening date of September 6 approached, Zhou urgently asked for help from John DeFrancis, the conference organiser and an eminent Sinologist. Fortunately, DeFrancis realised that a conference without representatives from China would be meaningless; he moved quickly to obtain funding for the five invitees, plus living expenses for them. Thanks to him, Zhou and his colleagues were able to board their aircraft and take part.

The conference was a success. It was, oddly, conducted in English, even though the subject was Chinese and all the delegates were scholars of the language. Zhou had sent them his speech in advance; he spoke about Pinyin, the standardisation of Chinese and modernisation of characters.

The Taiwan delegates spoke of the successful promotion of Mandarin on the island – at the end of World War Two in 1945, after 50 years of Japanese colonialism, Taiwanese and Japanese were the languages most widely spoken there. "I spoke of the situation in the mainland. Neither criticised the other," wrote Zhou.

His speech was later published in prestigious Chinese and English magazines. Relations between the scholars of the two sides were courteous. One outcome of the meeting was an increased recognition of Pinyin within Taiwan; this was no simple matter, given the political baggage attached to it.

The problem with Bopomofo, the Taiwan transcription system, was that it is like shorthand and is not written in Roman letters; you must first learn the shorthand before you can use it. So, it was difficult for foreigners. After the meeting, Taiwan revised Bopomofo to bring it closer to Pinyin. This made Zhou very happy. For political reasons, Taiwan could not adopt Pinyin outright.

Before China's open-door and reform policies, foreigners learning Mandarin went to Taiwan; it had a wide diversity of public and private institutions teaching it with traditional characters. From the early 1980s, an increasing number of foreigners went to the mainland to study Mandarin, with simplified characters.

Zhou observed that schools in Hong Kong and Taiwan emphasised the teaching of English, equipping their students well for international exchanges and meetings. The mainland was far behind. His foreign friends told him that China should pay more attention to English education.

While he was in Hawaii, Zhou had the good luck to be invited to another international meeting on "Pacific Basin Languages and Cultures" in Honolulu. That he was asked to attend a conference outside his area of expertise tells us of his growing reputation. He threw himself into it with his customary enthusiasm, meeting professors of Hawaiian culture and language and visiting museums devoted to it.

What he discovered was that, under the overwhelming pressure of migrants from Asia and the U.S. mainland and the dominance of English, Hawaii's culture and language were disappearing. Even local Hawaiians did not want to speak their ancestral language. Many residents were Japanese and wealthy Chinese attracted by Hawaii's climate and beautiful setting. In 1898, the United States had overthrown the independent kingdom of Hawaii and annexed it.

Translating *Encyclopaedia Britannica*

After paramount leader Deng Xiaoping's historic visit to the United States in early 1979, relations between the two countries warmed after 30 years of hostility. This led to many good outcomes. One was a Chinese translation of *Encyclopaedia Britannica* (EB).

In November 1978, the China Encyclopaedia Publishing Company was established. In August 1980, with official approval, it signed an agreement with the American publishers of EB to work together to publish a Chinese version. The two sides set up a six-member publishing committee, with three members from each country. Zhou was one of the three Chinese.

EB is the oldest English-language general encyclopaedia, first published in Edinburgh, Scotland in 1768. Since then, it has relied on both outside experts and its own editors to write the entries. They are fact-checked, edited and copy-edited by EB editors, to ensure readability and accuracy. In 1920, American retail giant Sears, Roebuck & Co bought the company. In 1941, it gave the firm to the University of Chicago; the city became its headquarters. In 2010, it printed 32 volumes of its traditional hard-bound volumes for the last time. Since then, it has published its content entirely online.

The amount of work involved in the translation was enormous. "Our conditions were too good," said Zhou. "Because we started work in 1980, many university professors had just returned from their May Seventh Cadre Schools and had no work to do. We could invite them to help us translate."

So they hired 500 professors and experts from Beijing, Tianjin, Shanghai and elsewhere. "They were delighted to do the translation. They felt the work was very meaningful. The fee was very little – but money was very tight. If we did it today (2010), it would not be so simple."

The first Chinese version was published in 1985, in 10 volumes, with more than 7,000 topics, 5,000 maps and a total of 24 million characters. Since then, the Chinese version has continued to expand, reaching 20 volumes in 2007. Zhou was an ideal person for this project. His nickname was Zhou Baike, Encyclopaedia Zhou. Interested in everything, he had a broad knowledge of many subjects, in addition to his specialties of economics and linguistics. In part through his work on the Education Committee of the CPPCC, he also had a wide range of contacts in the academic community.

By 1984, Zhou and his two fellow editors had completed more than half the work. To reward them for their efforts, the American publisher invited them to come to the United States. Always the scholar, Zhou wanted to give lectures as well as take a holiday. He sent seven talks to the publisher; he circulated them to American universities, asking if they would like to host one. On October 10, Zhou and his wife boarded a plane for San Francisco. It was his first trip to the mainland U.S. since 1946.

After their arrival, his wife stayed with Zhang Yuanhe, one of her sisters who lived in San Francisco. Four days later Zhou went to Santa Barbara, a wealthy city on the Pacific Coast 520 kilometres to the south and home to Frank Bray Gibney, vice chairman of the Board of Editors at EB and of the Chinese translation. Gibney was an accomplished journalist, editor, writer and scholar; he learnt Japanese at an elite Naval school and worked in Naval Intelligence, interrogating Japanese prisoners of war and officers. He wrote six books on Japan and East Asia.

Gibney treated Zhou like a VIP – putting him up in a high-class hotel with a large garden and a gift of papaya and other fruits every day. China was still emerging from the poverty and scarcity of the Maoist era; Zhou was dazzled.

He asked to see the home of the Mexican lady who cleaned his room. She had four rooms in two storeys with a small garden. "It was much more beautiful than my apartment in Beijing. It would take 10-20 years for a high-level intellectual in China to achieve the living standard of a chamber maid here," he remarked.

He gave a talk at the University of California at Santa Barbara on "The Story of the Alphabet". The audience of teachers and students were fascinated.

Gibney invited Zhou and 20 other guests to dinner at his luxurious home in Santa Barbara, with a large garden. Zhou found himself sitting next to a Hollywood film star and a member of the Cabinet from Washington. He learnt that the city's residents included a sister of the Shah of Iran and that President Ronald Reagan went there every month for a holiday. It was another planet from his Beijing world of a tiny study covered with books and political meetings in crowded rooms full of cigarette smoke.

He reflected on how the U.S. had changed since he had lived there 40 years before. "Before, the U.S. developed large cities. Now everyone has a car and the road network is extensive. Many people live in small and medium-size cities, and air travel is very convenient. Distance means nothing. This was a great change."

The next stop was Chicago, where the U.S. publishing company had its headquarters; Frank Gibney flew with him. Zhou stayed in a similarly high-class hotel, where rooms cost US$200 a night. His suite had two rooms; the telephone, with three extensions, had been made in Taiwan. He gave a lecture at Chicago University on the history of Chinese characters. It was well received by the audience of faculty and students. Then he went to Urbana-Champaign to give a lecture at the University of Illinois. For his audience, the main interest was the homonyms – characters with the same sound. Pinyin helped but did not solve this problem.

Next was Ann Arbor, Michigan. No lectures here, but five days of relaxation with a cousin of his brother-in-law named Tu Guo. Having stayed on in the U.S. after his graduate studies, Tu worked as an engineer in a large machinery manufacturing company; he rose to become its president. At that time, few Chinese in the U.S. were so successful.

It was a chance for Zhou to see the lifestyle of a wealthy American family. Tu's wife was a Chinese who had also studied in the U.S.; her father had been a Minister of Communications in the Nationalist government. The family lived in a spacious Western-style home with a garden and a large dog; it served both to protect the house and as a companion to the family. The house was well equipped with many conveniences.

After returning from a busy day at work, Tu did household chores; on Sunday mornings, he rose early to mow the lawn. Zhou saw the kind of life he might have had if he had stayed in the U.S. after 1949 and not returned to China. With his social and professional skills, he could have risen to a high position in business or finance and bought a similarly large house. Tu's wife complained that, at school, their children were bullied by both their white and black classmates.

In November, Zhou moved on to New York, where he had lived in 1946 as the representative of Xinhua Bank. There he met three close Chinese friends; they had set up in New York "The Society to Promote Reform of Chinese Characters" and were its most important members. The three occupied different places in society.

One was a technician. Zhou stayed in his house, with four rooms in three storeys and a small garden. The house was in a black district. The second was an interpreter with the United Nations earning US$100,000 a year, a sum unimaginable for a mainland Chinese at that time. The third was head of the Chinese Section of the New York Public Library.

During his posting to the city in 1946, Zhou had spent most of his evenings at the library. He was delighted to find that the building had scarcely changed in 38 years, except for the addition of new sections paid for by donations from wealthy individuals. It was a building of historical and artistic value. The big change was a new computer system, through which readers were able to trace books; the card index files were still there, but few people used them.

American universities had linked their computer systems, to facilitate readers looking for works. Zhou discovered that, to find books in Chinese, these universities used the Wade Giles system of Taiwan, not Pinyin. "There were two reasons for this. The vast majority of Library Directors came from Taiwan; they opposed Pinyin. Americans did not want to do Library Studies because jobs in the field paid only modest salaries, but Taiwan people were happy to do (these studies). The second reason was that the cost of switching to Pinyin and the time involved were substantial, so few wanted to do this." As we described in the previous chapter, the U.S. libraries switched to Pinyin in 2000.

In addition, Zhou met a former classmate who had started out in import and export between China and the U.S. and branched out to other countries. He specialised in the export of American machinery. He set up his own company and opened a lavish office in the World Trade Center; the two towers were the highest buildings in the U.S. at that time. "He owned 100 per cent of the firm and was one of the most successful of my friends in the U.S."

He invited Zhou to a lunch at an expensive restaurant in one of the towers. It is indeed impressive that, despite being unable to leave China for 34 years and restricted in his communications with the outside world for most of that time, Zhou had maintained such close friendships across the Pacific Ocean.

On Zhou's schedule of lectures was the United Nations, which had one of the largest concentrations of interpreters in the world. His audience was the "United Nations Workers Language Society"; he spoke about the history of the Chinese language. This fascinated his audience and stimulated many questions. The director of the society, a French lady, praised Zhou, saying that the topics chosen by many foreign speakers often did not interest their members; but his topic did. "At that time, Sino-U.S. relations were good and China had only recently entered the United Nations. So the atmosphere was to welcome China. I felt very happy."

He also gave two talks to smaller audiences at the United Nations. One was to members of the Chinese department; their job was to translate from Chinese into other languages and vice versa. It was the largest single language department in the United Nations. Despite the high quality of its staff, it was the least efficient department; this was due to the difficulty of Chinese and the fact that so many characters had the same sound. Those working there earned US$100,000 a year and enjoyed other benefits. But few in the department came from the mainland, because the government sent only a limited number.

Zhou reflected on the irony that, except when he was addressing an entirely Chinese audience, all his talks in the U.S. were in English, even when he was talking to professors of Mandarin. "The number of Chinese speakers is very large but it is the language of one race and one country. It is not global. This is a very big problem." By contrast, the five other official languages of the United Nations were spoken in many countries.

Zhou was fortunate to be in New York the evening of the presidential election on November 7 between Republican Ronald Reagan and Democrat Walter Mondale. There were four other candidates, including one from the Communist Party USA. Zhou was fascinated to see how people voted; they had to prove their eligibility, were given a voting paper and then put this into a machine. He and a friend spent a long time watching this procedure. Then they returned to his friend's home, where they watched the results come in on his computer. "This voting system via computer is common now. But, at that time, the U.S. was the only country to use it. It was fast and accurate." Reagan won by a landslide, with 58.8 per cent of the vote, against 40.6 per cent for Mondale; Reagan won 49 of the 50 states.

Another thing that struck Zhou during his visit was the relative decline of New York, compared to the city he had known in the 1940s. While it had retained its economic status, it was short of money; this led to many problems, such as insufficient collection of rubbish and worsening crime.

A former classmate, retired, told Zhou that he went for a walk every morning and evening. "I cannot take no money and do not take too much. I carry US$20. When a black threatens me, I give him the US$20 and say that I am a retired person and do not have much money," he said.

Similarly, Zhou spoke to a nurse who lived on her own. She said that, when she went to work, she left a US$20-note on the table of her reception room for a robber. Next was a piece of paper saying: "If you need money, please take this. I am a nurse and do not have money."

In November, Zhou went to stay with a younger sister of his wife and her family in New Haven, Connecticut. Her husband was Hans Frankel, a Sinologist and a professor at Yale University. At his invitation, Zhou gave a lecture at Yale. During World War Two, the university had developed the "Yale System" of Romanised Chinese, close to Zhou's Pinyin.

He called his wife, who had been staying with another sister in San Francisco; she took a plane to join him and the family. Zhou had no more lectures, so he was able to take it easy. Ever diligent, Zhou read many books and magazines not available to him in China; he learnt many new things. One was a detailed account of how Singapore had become independent in 1965, leaving Malaysia of which it had been a part since the country was founded in August 1957. Zhou marvelled at the rapid pace of Singapore's development since then, especially during the more than 10 years when China was plunged into the tragedy of the Cultural Revolution from 1966.

Mrs Frankel was Zhang Chonghe, younger sister of Zhou's wife. In January 1949, she and her husband had left Beijing to move to the United States. First, they lived in California, where her husband taught at Berkeley. After emigrating to the U.S., they adopted two children. In 1961, they moved to New Haven, where he accepted a professorship at Yale University.

Chonghe taught calligraphy in the Arts faculty. She was an accomplished and energetic performer of Kunqu. She sang, played the flute and taught and performed in over 20 universities across the United States as well as in France, Hong Kong and Taiwan.

"Unlike many migrants who moved overseas, Chonghe did not change her profession after moving to the United States," wrote Kang-I Sun Chang, Professor of East Asian Languages and Literature at Yale University. "For many years, she continued to promote the calligraphy and Kunqu she dearly loved. In a completely strange environment, she exercised a special influence (in Kunqu)." She also taught Kunqu to students at her home in

Connecticut, well into her 90s.

Zhou had one fascinating day in New York City. Roxane Witke, a former student of Professor Frankel and history professor at Columbia University, invited Zhou and other family members to a meal. She told them the remarkable story of how she had written *Comrade Chiang Ch'ing* (Jiang Qing in Pinyin), a biography of the wife of Chairman Mao and one of the Gang of Four; it was published by Weidenfeld & Nicolson in London in 1977.

She had gone to Beijing and befriended Jiang Qing, with the help of Premier Zhou Enlai. During the summer of 1972, Jiang, then at the top of her power, gave her a week of interviews. Learning that Witke liked swimming, Jiang invited her to a large pool in Shanghai – and Witke found that the two of them were the only bathers in this exclusive place.

After reports about her book appeared in the media, China's ambassador in Washington, Huang Hua, invited her for a meal. He proposed that, before she submit it to a publisher, the embassy buy the text from her, for more than US$1 million, a fortune for a history professor on a modest income. She refused. The book was published after Jiang's arrest in October 1976. It was mainly the words of Jiang herself; it was banned in China.

Since Zhou could not take the book to Beijing, he read it at the home of his sister-in-law before his return. "After I finished, I had no special feeling. Of course, people outside did not know what Jiang had to say about herself. That was something. I did not consider it exceptional."

Having lived through the Cultural Revolution in China, Zhou knew well the life and ideas of Jiang Qing. It would be more interesting for a non-

Chinese, ignorant of the mainland. After her arrest, Jiang was put on trial for being part of the "Lin Biao and Jiang Qing Counter-Revolutionary Cliques" and sentenced to death. In 1983, the sentence was commuted to life imprisonment. After being released for medical treatment, she hanged herself in Beijing on May 14, 1991.

Zhou and his wife had one more adventure before returning to Beijing. They flew to San Francisco and visited the gambling city of Reno, Nevada, beloved by many Chinese. He bought US$25 worth of chips – and lost the money in no time. They went back to Beijing in February 1985.

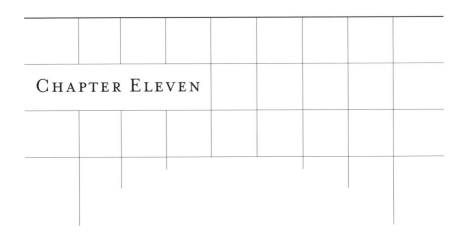

CHAPTER ELEVEN

Pinyin Brings Mandarin To The Information Superhighway

The greatest contribution of Pinyin has been to enable tens of millions of Chinese to become literate in their own language, as well as help millions of foreigners to learn it. It has also played a major role in taking Chinese into the electronic age and turning it into the second most used language on the Internet, after English. This is something beyond the imagination of Hu Shih, Chao Yuenren and other language reformers of the early decades of the 20th century.

Initially, there were two main methods of inputting Chinese into a typewriter or computer. One uses a phonetic system, of which Pinyin is the most common. The other is to use one based on the roots or radicals of the characters. Using Pinyin, the writer types the spelling and the computer offers the character matching the spelling; since many characters have the same spelling, this process is not so fast, as you spend time selecting the character.

To speed this up, Pinyin software often has a predictive text function, suggesting characters based on context or common usage. Writers can also create a dictionary of the characters and phrases they use most frequently. Pinyin is the method most foreigners – including me – use to write Mandarin. My computer offers both simplified and traditional characters. If I am writing to someone in the mainland, I use simplified ones. If I am writing to Chinese elsewhere, then I use traditional ones.

The most common method using the root of characters is Wubi (Five Pens). This divides the keyboard into five key areas and creates its own shorthand. Nearly every character can be written with at most four keystrokes; most can be written with less. The keyboard uses the standard QWERTY format, with additional labels on each key. Experienced users can work very fast, with up to 160 characters a minute. To reach this level, they must spend time learning the system. Pinyin and Wubi are the methods most commonly used in mainland China.

In Taiwan, the popular method of inputting into the computer is Bopomofo, the system children study at school to learn characters. It is based on the pronunciation of Mandarin. Another method of inputting has arrived with the development of touch-screen technology. A user can simply write the characters on the screen on his mobile, iPhone or other device; he does not need to learn Pinyin, Wubi or Bopomofo. The software is often able to understand the character you are writing, so you do not have to write all the strokes.

Another method is rapidly gaining users – voice input. You speak to a microphone and the computer or mobile phone will turn your language into words; this is faster and more convenient than Pinyin. This means that even those who cannot read the characters can use voice input to chat with friends on WeChat, China's most popular messaging system.

Helen Wu, who lives in Shanghai, said that her teenage daughter and her classmates were increasingly using voice input. "They hate writing compositions and use voice input to write them," she said. "I forbid my daughter from doing so. Too much convenience will prevent her from deep reflection."

These different methods have become popular and widespread today. They would not exist without the ingenuity and hard work of thousands of Chinese engineers in the mainland, Taiwan, Hong Kong and elsewhere. Bringing Mandarin into the computer age has been essential to make it one of the most important global languages on the Net, for high technology, science, medicine, engineering and other important uses.

Chinese have embraced the Internet with great enthusiasm. China joined the Internet age on September 20, 1987, when Qian Tianbai, a professor in Beijing, sent the country's first e-mail; it was aptly called "Crossing the Great Wall to Join the World". In 1988, Chinese universities linked with Europe and North America. On April 20, 1994, China opened a 64K international dedicated circuit to the Internet through Sprint of the U.S., linking it to the Internet.

In September that year, it started work on its own Internet, the Chinanet, primarily aimed at colleges, universities and middle schools. In December 1996, the country's first websites opened, in Shanghai, Guangdong and elsewhere. By the end of 1997, China had more than 290,000 personal computers and 620,000 Internet users. By the end of 2000, the number of users rose to 22.5 million and, by the end of 2005, 111 million. Since 2008, it has been the country in the world with the largest single number of Internet users. According to Internet World Stats, as of March 31, 2020, English ranked first in the world with 1.186 billion users, followed by Chinese with 888 million and Spanish third with 364 million.

When he was creating Pinyin in the 1950s, Zhou could not have imagined how useful his invention would turn out to be. The millions who use Pinyin every day to access the Net owe him a big debt of gratitude.

Pinyin in Taiwan

While Pinyin has conquered most of the world, one place where it has limited use is Taiwan. The reasons for this are many and complex. At the end of World War Two in August 1945, Japan handed Taiwan back to China after 50 years of colonial rule. Nationalist officials sent to manage the island discovered to their surprise that Mandarin was spoken by only a minority of people, mostly intellectuals, among the population of six million.

The main language was Taiwanese, spoken by 70 per cent, followed by Hakka, about 15 per cent, and the rest one of many Aboriginal languages. The island's educated elite, including professors, lawyers, doctors, bankers and civil servants, also spoke Japanese, the official language during the 50-year colonial period.

During the 19th century, Western Presbyterian missionaries had created a Romanised form of the Taiwanese language. They did this after discovering that few of those listening to their sermons could read the Bible written in Chinese characters. They wrote a dictionary of this new language, which became the officially accepted form of written Taiwanese, known as Peh-oe-ji. Books, newspapers and magazines were published in the language, and people wrote letters in it. As a Romanised form of a Chinese language, it was the same idea as the xin wenzi (New Characters) developed by Qu Qiubai and his Soviet associates in the 1920s and 1930s. A monthly newspaper using this Taiwanese language was published from 1885 until 1969.

After his defeat by the Communists in 1949, President Chiang Kaishek moved with 1.2 million soldiers, civil servants and other citizens from the mainland to Taiwan. Like Mao and Zhou Enlai in Beijing, Chiang made language reform a priority of his new administration in Taipei. He determined to make Mandarin the sole national language, written and spoken. He saw it as a unifying force in a country where people spoke different languages; it was also a common language between Taiwan and the mainland. Chiang saw Hakka and Taiwanese as dialects of Chinese. Mandarin was the only language in schools and universities, the government, army and police and the media.

This policy was opposed by many Taiwan people, angry that the languages they spoke had no official status and the fact this gave the Mandarin-speaking arrivals from the mainland an unfair advantage in many sectors of life. The main instrument of this Mandarin-only policy was the schools. All the teachers and students might be Taiwanese-speaking but they could use only Mandarin, even in the playground. Naughty children who broke the rule were scolded by their teacher; they hang around their neck a board reading "I must speak Mandarin".

At home and in their companies, shops and restaurants, Taiwanese spoke their own language; they put satellite dishes on the roof to enable them to watch Japanese television. It was illegal but tolerated; it was called "the fifth channel", in addition to the four official ones, broadcasting only in Mandarin.

In promoting Mandarin, Chiang was following the same policy as Mao on the mainland. Both have succeeded. Today the large majority of people in Taiwan speak, read and write Mandarin, as they do on the mainland. Taiwan has one of the highest literacy rates in the world, 98.7 per cent, up from 78 per cent in 1970.

But the two sides diverged after Beijing announced in 1958 – at the same time as Pinyin – a new vocabulary of simplified characters. Taiwan considered this a "bastardisation" of the written language. Until today, it has proudly continued to use the traditional characters. Taiwan people recognise that simplified characters have helped to reduce illiteracy as they are easier to learn than traditional ones. But they say that, while they retain the sound, the new forms destroy the original design of the characters. For example, "love" is "ai": in traditional form 愛 , in simplified 爱 – but it has lost 心 , xin, the heart. To fly is "fei": in traditional 飛 , in simplified 飞 – but it has only one wing. Taiwan people watched the 10 years of the Cultural Revolution from 1966 to 1976: Red Guards destroyed books, temples, churches, art works and cultural treasures. What authority did the mainland have to speak in matters of Chinese culture and history?

The Nationalist government did not adopt Pinyin. It had brought from the mainland the Wade-Giles system for writing in the Roman alphabet and the Bopomofo phonetic system for students to learn the characters. These were widely used and accepted by the population. So, it saw no need to use Pinyin.

When children in Taiwan go into primary school, they first learn Bopomofo – as those in the mainland learn Pinyin – and then they learn the characters. Similarly, to type on the computer and access the Internet, most Taiwanese people use a keyboard based on the Bopomofo system. "Its symbols are based on ancient Chinese characters," said Liang Chunhsiu, a Taiwan secondary school teacher. "Why should we use one based on Roman letters?" Students learning Taiwanese, Hakka or Aboriginal languages first learn Tongyong Pinyin, a romanisation system very similar to Pinyin.

Hostile relations between the two sides make it difficult, if not impossible, to accept ideas from the other. In 1998, Taipei city, under the control of the opposition Democratic Progressive Party (DPP), adopted Tongyong Pinyin. But, in 1999, the Legislative Yuan, controlled by the ruling Nationalist Party, announced that Taiwan should adopt Hanyu Pinyin. So there were two competing systems. Each party used the system in the areas it controlled.

Finally, in January 2009, the government declared that Hanyu Pinyin would be adopted as the official system for Chinese romanisation. "This brings us in line with international standards," said Chen Hsuehyu, executive secretary of the Taiwan education ministry's National Language Centre. "Foreigners reading signs won't say 'I don't understand' anymore."

But it is not uniformly practised; the main cities retain their Wade-Giles romanisations, such as Taipei, Kaohsiung and Hsinchu, as they are known around the world, and not the Pinyin forms of Taibei, Gaoxiong and Xinzhu. A Taiwanese can choose the romanisation of his or her name in their passport. The president calls herself Tsai Ingwen, not Cai Yingwen; in the Taiwanese language, she is Chhoa Engbun.

Pinyin remains a sensitive issue, not because of the merits of the system itself, which few can challenge, but because it comes from the mainland.

Since a DPP president took power in 2016, relations with China have deteriorated. The People's Liberation Army conducts regular air and naval exercises close to the island and Beijing has not ruled out reunification by force. All this fuels hostility toward the mainland – and things connected to it. On the other hand, millions of Taiwanese people have studied, worked and lived in the mainland; they watch mainland films, operas and television programmes. So they have become accustomed to Pinyin, even if they do not know the name of Zhou Youguang.

During his life, Zhou never visited Taiwan. But he knew many Taiwanese people, especially scholars in the linguistic field and friends and colleagues who went there with Chiang Kaishek in 1949.

In Hong Kong also, he is not well known. The main language used in Hong Kong schools is Cantonese. Students learn the characters directly, without using a Romanised system. But, in their Mandarin classes, they learn Pinyin.

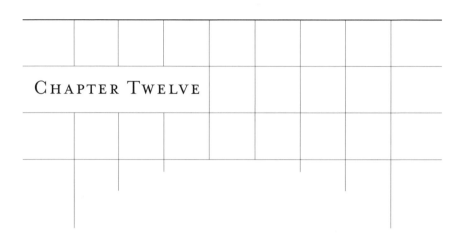

CHAPTER TWELVE

"God Is Too Busy, He Has Forgotten Me" – Final Years

On December 31, 1991, Zhou officially retired, two weeks before his 85th birthday. The CCRC gave him a retirement certificate, which he discovered was dated December 1988. "They were too embarrassed to force me to leave. So, they waited until I left on my own – a period of three years."

At that age, most people would have chosen a life of long lunches, chess, card and mahjong games with their family and friends, tourism, reading novels and naps after lunch. But Zhou did the opposite. During his remaining 26 years, he wrote more than 20 books; he averaged one article a month and one book every three years, published in the mainland, Hong Kong and Taiwan. They covered a wide range of subjects, including history, literature, anthropology and poetry as well as his specialties of linguistics and economics.

He lectured at Peking University and Renmin University of China . As he grew older, he became increasingly critical of the Communist Party, the Soviet model it had adopted, its version of history and its treatment of intellectuals like him. These subjects could not be published in the mainland. But the authorities left him alone, probably because of his high status and advanced age.

He said in an interview in April 2010: "When you are young, you are naïve and follow blindly. In old age, you start to explore the truth. I am 105 and could die tomorrow. It is no problem to say the wrong thing. Others who write articles must be careful."

In addition to a drive to write, what sustained him was a constant stream of family, friends and visitors, Chinese and foreign. He was delighted to see them and exchange news and opinions. They marvelled at his smile, jokes, optimism, stamina, encyclopaedic knowledge and willingness to speak the truth. Foreign journalists sought him out for this reason. He was one of the few people in Beijing willing to say publicly that the Emperor had no clothes.

How did he live so long? He did not drink or smoke, nor eat health supplements. "Do not get angry. Be more tolerant," he said. "I am an optimist. Whatever difficulties you face, see the good aspects. The bad aspects will slowly pass … My best tonic is remaining curious and continuing to learn."

He was also sustained by humility and contentment with a simple life. "My life is very ordinary, with no special value. I am an average person … God is too busy and has forgotten me. The Christians have a saying, that God welcomes His children to go to Heaven. I am a child whom He does not like, so I have not gone."

When visitors praised him as "the Father of Pinyin", he would modestly answer that he was "the Son of Pinyin" or one of its main creators. "It is the result of a long tradition from the later years of the Qing dynasty down to today. But we restudied the problem, revisited it and made it more perfect." He praised the hard work of his colleagues at the CCRC. But the official record shows that, within the CCRC, he was the director of the Pinyin Research Department.

The institution offered him a larger and more comfortable apartment in Fangzhuang, on the outskirts of Beijing, but he declined to go. His material possessions were limited. During their life, he and his wife had lost everything twice – once after World War Two and once on returning to their Beijing home in 1972, when not one scrap of paper remained.

"That is fine – no assets and no burden. I feel very relaxed," he said. "There is a Buddhist saying – if you attach too much importance to items outside the body, your spirit will suffer ... During the Cultural Revolution, we were sent to Ningxia. Everyone believed that we would not come back and became very depressed. I felt it was fine. If I had not gone, I would never have known such a place. So, I did not get angry. Getting angry achieves nothing. I have lived through so many difficulties and experiences."

Zhou and his wife lived in a small third-floor apartment in a grey block in central Beijing in Chaonei Dajie Guaibang Hutong (Crutch Stick Alley); it had less than 50 square metres. It was, ironically, on the site where his "cow pen" had been during the Cultural Revolution. The "cow pen" – garage – was demolished and an apartment block built there. Facing east, his apartment had four rooms. Two had a space of 14-15 square metres, with a room of eight square metres for the maid and one of 9.4 square metres for his study.

His study had a table, two chairs, a tea table and a bookcase up to the ceiling. It contained a red mahogany cabinet for documents, the only item that survived from their former home before he was sent to Ningxia. He sat on one of the chairs and his beloved wife on the other; they drank black tea and coffee. After she passed away in August 2002, at the age of 93, he swapped one of the chairs for a sofa. From then on, he slept on the sofa and did not return again to their bedroom to sleep.

He hung on the walls pictures of his wife at different stages of her life and of other family members – but not images of the government leaders he had met. The study had a window one metre high, facing north; outside was a paulownia tree. Each morning many birds gathered on the tree and sang – Zhou's morning alarm clock.

Each day he read at least five newspapers, including some brought from

Hong Kong and overseas. His regular reading included the American magazines *Time* and *Newsweek* and the Chinese-language *World Affairs magazine*; his many visitors also brought him material to read.

In his later years, Zhou used a hearing aid, a magnifying glass and walking stick and was looked after by a live-in maid. He drank Starbucks coffee and listened to modern Western music.

His niece, Mao Xiaoyuan, who lived and worked in Beijing, saw a great deal of him during this period. In October 2015, she wrote an article to mark his 110th birthday. "Those many years with Uncle made me feel deeply that I was a drop of water and Uncle was a great ocean."

She described his energy and enthusiasm for life and learning, how he collected information from China and overseas and had friends with a similar wide range of interests. "The real scholar does not leave his home, knowledge comes to him from the sky," she wrote.

She described outings with him to parks and other scenic places in Beijing. In December 2014, he addressed a university audience of several hundred on "Comparative Literature"; he spoke for an hour without notes, answered questions for 90 minutes and then had his picture taken with many of the audience. In the summer of 2003, the family drove him to the seaside resort of Beidaihe, where Party leaders go in August for meetings and swimming. When the manager of the guesthouse saw Zhou and recognised him as the Father of Pinyin, he insisted on giving him three days of free board and meals. Zhou, 98, took the opportunity to go swimming.

Until 1988, Zhou wrote articles by hand. Then he received from the Japanese company Sharp one of the earliest Chinese-language typewriters it had developed for the China market. At that time, electronic ways of writing Chinese were undeveloped. Sharp asked Zhou for his help; he gave them detailed advice on Pinyin. "After they developed it in 1988, they gave me one. The price then was 5,500 yuan, which many Chinese could not afford. It is only for writing, not like the computers people buy today, on which you can play games. After I acquired this, my speed of writing increased."

He treasured his new acquisition. Each evening he wrapped it up and stored it carefully; next morning he unwrapped it and set to work with enthusiasm. He taught his wife how to use it. Her focus was as director of the Kunqu Research Association.

The new Sharp greatly increased his output. During the 1980s, he published or republished four books. In the 1990s, it was 13; by March 2010, the number had reached 22. The book which made the greatest impact and sold the largest number of copies was *An Outline of Chinese Character Reform*. It was first published in China in 1961, then republished in 1964 and 1979. It was also published in Japanese, German and English. "In the 1950s, there were fees for writing articles and none in the Cultural Revolution. The fees (in the 2000s) are very low, little changed from the 1950s, while prices have increased a dozen-fold. If you lived off fees from articles, you would die of hunger!"

Passing of Beloved Wife and Son

During his retirement, the event that hurt him the most deeply was the passing of his beloved wife on August 14, 2002, at the age of 93. They had been married for 70 years. "The night before, a friend came to take pictures. The next day she died and left this world. Her heart was not very good. At 93, we should say that death is normal. Her passing was a bolt from the blue. I had never imagined one day when we would not be together. Such a blow suffocated me, but I had no alternative but to accept the law of nature ... I thought that, since I was four years older than her, I would go sooner. So, I gave the valuable items in our house to our granddaughter. I did not imagine that I would live until now," he said in an interview in 2010. "People say that, when you get old, each day you live is one less. This is completely wrong. Each day you live is one more."

Granddaughter Zhou Heqing said that, after the death of her grandmother, she, her husband and son flew from their home in California to see Zhou. "Late that night I sat with Grandfather in his study. It was the first time I have seen him with redness around his eyes. He said that Grandmother's passing was too sudden and no-one had expected it. 'Her body was always weak but her determination to live was so strong and she was full of life. You must not worry. I know what to do. I hope not to give you trouble during this time.' Normally, Grandmother sat next to him and his computer. But she would not be coming back."

He and his wife were extremely close. On her 80th birthday in 1989, he wrote her a love poem and gave her a copy of *Romeo & Juliet* in English. She was an eminent scholar of Kunqu, one of the oldest forms of Chinese opera. She liked to drink green tea, listen to traditional Chinese music and sing Kunqu. Zhou preferred coffee and black tea and Western music.

On August 24, friends and family took her ashes to a hill in the western suburbs of Beijing and planted a maple tree, next to which they scattered the ashes. Zhou remained in quiet mourning at home and did not go with the group for the event. Later that day he joined his friends for a meal at a vegetarian restaurant. He signed the visitors' book and took photographs with his friends. In October that year, family and friends produced a magazine in memorial of her. In 2004, her *Kunqu Diary* was published.

After her passing, Zhou himself twice went to hospital with a serious illness but recovered. He stopped using the bedroom and slept on the sofa in the study. In a later interview looking back on their marriage, he said: "My view is that a couple need not only love but also respect. We two both drank tea and coffee, treating the other with courtesy. With love and respect, marriage will be more complete. We had 70 happy years of marriage."

Another heavy blow came in January 2015. His son Zhou Xiaoping went to hospital for a major operation on his stomach. After the procedure, he went to stay with his father on January 10. After dinner, that evening, the two sang together. Youguang sang in English the school song of St. John's University and, in French, La Marseillaise, France's national anthem. It was the first time in many years the two had sung together. Father was 109 and Son was 80.

Xiaoping stayed there for five-to-six days before going home. In the early morning of January 22, he passed away. He had a distinguished career as a meteorologist. After studying for three years in the Soviet Union, he returned to China in 1962 and joined the Chinese Academy of Sciences as a researcher.

He held senior positions at the Academy and won awards before his retirement in 1999. The funeral was held on January 26; Zhou's niece Mao Xiaoyuan went to stay with Youguang. She recalled: "One evening, his amah pushed his wheelchair next to my bed. Uncle spoke at great length, talking of the life of Xiaoping and saying that this was the law of nature. His words gave us strength."

After his son's passing, Zhou went himself to hospital and, fearing that his life was at risk, wrote his testament three times. But, on each occasion, he returned home.

The Beijing government did not include Zhou in its policy-making, even on linguistics. In 2010, the Ministry of Education issued a revised version of commonly used characters. "They did not consult me. They probably thought I was too old. I read about it in the newspapers. Many people opposed this. They wanted to increase the number of characters from 7,000 to 8,300. They did not explain clearly. I said that it was wrong. We cannot increase from 7,000. Of the 7,000 we researched, 400 had problems and we were not certain to use them. Many were dialects or rarely used. My view is that you can only reduce the 7,000, not increase them. Characters are a bottomless pit."

Breaking Taboos

In his writings and interviews between 1992 and his passing, Zhou broadened his interests far beyond the linguistics on which he had concentrated in his professional life. Of the 49 books he published during his life, he wrote 33 during this period. They cover a wide range of subjects, including world history, culture, anthropology and society.

Zhou wrote about topics taboo in China; these included a critical examination of its history, especially since 1949, its treatment of intellectuals, the role of Mao Zedong, and June 4, 1989. Zhou was one of few Chinese intellectuals willing to go on the record on these subjects. This was one reason so many people, Chinese and foreign, wanted to meet him. It could be dangerous to speak publicly on such topics; the government probably left him alone because of his high status and advanced age and the fact that it could censor his words published in the mainland.

Asked about his decision to return to China in 1949, he said: "History misled us. But I do not regret coming home. Despite the traumas we witnessed and endured, this was where I belong … At that time, we all felt that China had hope. The construction of China was waiting for us. I had studied economics for so many years. I thought that what China most lacked at that time was economic construction. So, I went back home to build the economy."

In an interview with *Der Spiegel* published in February 2014, he said: "I have only one regret in my life, that the Communist Party did not keep its promise and China has no democracy. What Mao Zedong did was to leave China in a complete mess."

In an interview in April 2011, he said that, during the anti-Japanese war in Chongqing, he attended a monthly meeting of 20 people, including Zhou Enlai. "Each time Zhou Enlai said that the Communist Party advocated democracy. At that time, we all despised the dictatorship of the Kuomintang. Later I organised a group of five important people to visit Mao Zedong in Yan'an. Mao said that, to govern China for a long period, democracy was the way."

Zhou Youguang said that democracy was an inevitable path for China. "It is not something discovered by some countries or a patent. It is the result of 3,000 years of human experience. We (China) are running at the back. The good thing is that we have had reform and the open door. By the fastest way (to democracy), we will need 30 years, the slowest 150 years. The two biggest problems are income disparity and corruption. But democracy is no simple matter and we must not be rushed. There is a lot of work to do – first, we must establish laws. Since the start of the reform and open-door, we have passed some excellent laws, such as the property law. For democracy, we need two pre-conditions – we need to be open and improve the quality of thinking of the common people."

In 2010, he published *Shi Bei Ji, Collecting Shells – The Words of a 105-year-old Man to Awaken the World*. "I am 105, I will die tomorrow so that I can say the wrong thing," he said. It was published by Cosmos Books Ltd. The book expressed the anger and bitterness of thousands of intellectuals of his generation who felt that the Communist revolution had betrayed them and wasted their talents and their patriotism.

"Mao Zedong accepted everything from the Soviet Union. 'Reforming' intellectuals meant wiping them out. The Soviet Union sent intellectuals north of the Arctic Circle. By the time of (Nikita) Khruschev, half were dead. I myself belonged to the intellectuals to be wiped out. When we came back from the Cadre School in Ningxia (in 1972), the State Council summoned high-level intellectuals. They said that we were 'the dregs of society' and useless. For humanitarian reasons, we were being given food to eat. 'You go home and do not act or say anything improper. You have no work.' If it had not been for the Lin Biao incident, we would not have come back (to Beijing).

"Power does not allow people to speak the truth, only to tell lies. With the open door and reform policies in the 1980s, natural science and economics were introduced. To develop the economy, you need the study of economics. But many social sciences were not brought in, nor the study of education. Now we have meetings on education but they result in no change, because we cannot introduce reasonable educational theory. Political science and other social sciences also cannot be changed. In the future, they can change slowly. The Soviet Union did not even allow natural science, so it collapsed … The role of intellectuals in China is to pursue the road of science and democracy. Today we are walking on the path of science but not on the path of democracy. So, we are still a backward country."

History and Fake History

He also criticised the official version of history. "From what I learnt before, it was Hitler who launched World War Two by invading Poland. This is not correct. It was Germany and the Soviet Union, through a secret agreement to divide up Poland, who started the war."

But China had made progress. "When we translated *Encyclopaedia Britannica* in the 1980s, we came to a problem in describing the Korean War. We said that it was launched by the Americans, but the Americans said that it was the North Koreans who launched it. So, in the first edition, we did not include this section. In the second edition, in 1999, things were more relaxed. We agreed that it was the North Koreans who started the war.

"Previously, China said in its propaganda that it was the Communist Party that did most of the fighting against Japan in World War Two. Now we recognise that the battle area of the Kuomintang was large, it had many armies and fought for eight years right up to the end. Our history books are making progress."

He said that his granddaughter had studied for two years in Britain and two years in the United States. "She said that her studies went smoothly in both countries, but there was one thing she found very hard to accept. Foreign students had small-group discussions. When it came to China, the foreigners knew the subject, but the Chinese students did not.

"Today so much of what Chinese read is fake history, too much fake history. Of course, this should be revised, slowly. After the collapse of the

Soviet Union, Russia did a good thing – it opened its archives 24 hours a day. Every day people came to read them."

This has not happened in China. "China does not want you to have common knowledge. They themselves do not have common knowledge. People are not allowed to have common knowledge and that is the case today. How many books have I published? They remove the sections on democracy and do not ask my opinion. Even more ridiculous is a book published 60 years ago by Shen Congwen [his brother-in-law]. They removed from it the section on democracy. It is very sad."

"Soviet Union Caused the Greatest Harm to China"

On economics, he said in 2010 that there was no such thing as the "China miracle". "There is no miracle, only the rules of the market. Everyone today says that China has taken off. I do not believe so. We are still at a very low level. We must keep changing and raise the level of technology we import. Many university graduates cannot find work because the economy is not developed enough. Recently, newspapers have reported that factories cannot find workers because the wages are too low. That is a good sign – low-cost production will go elsewhere."

He praised the example of Singapore, which had become an independent state only in August 1965 and had built a stable and prosperous country, by following the market rules.

He was very critical of the Soviet Union. "In the 1930s, the Soviet Union was very powerful. The dictatorship of the Kuomintang was copied from the Soviet Union. It controlled both the Kuomintang and the Communist Party. It harmed us. Now the conclusion of everyone is that the misfortunes of China all came from the Soviet Union. It caused the greatest harm to China.

"The Soviet Union collapsed because it was against the rules of history. You can do that for a short time, but not a long time. The Socialism built by Stalin was fake. It was a deep dictatorship. After the Soviet Union collapsed, Russia returned to its imperial traditions and started capitalism again. The Soviet economy was a 'greenhouse'. As soon as you open the window, the cold wind will come in and the flowers will die. The Soviet planned economy was not competitive. Three days after the collapse of the Soviet Union, the red flag came down. Why was that? Soviet people did not speak, but they knew in their hearts. The army also knew – the deceit could not go on."

He said that, before 1949, the standard of Chinese banks was very high, not far behind those of the U.S. "At that time, we had already entered the era of cheques. Now many people do not know what a cheque is. Then, after 1949, we followed the Soviet model. The Soviet banking industry was terrible. But, in the last 20 years, we have made very fast progress [in banking]."

Tragedy of Soviet Model

He criticised many elements of the Soviet model which China adopted in 1949. One was the "stuffed-duck education", which requires students to memorise enormous amounts of data, in order to pass very competitive exams; they forget most of the data soon after.

"Chinese education has two major problems. One is the lack of academic freedom at the universities. They have no academic freedom. They are run by officials who bring absurd ideas into universities and run administrative departments, like professors. The other is the enormous amount of useless labour. Primary and secondary students work until very late and are exhausted. But they are wasting their time. If a VIP comes to the campus, students have to form lines and greet him. Whatever for?"

He preferred the liberal methods used in his schools in Changzhou and two universities in Shanghai; the course work was limited and the teachers encouraged the students to do their own research and interact with their classmates and the teachers. "If there is no independent thinking, there is no education. Education today has failed in this. It trains people to believe blindly in Marxism. You have no choice but to believe – this lacks independent thinking."

He also criticised Soviet "economics". "Previously, China only had Marxist political economics. There was no real economics. Communism denies capitalists, saying that they exploit workers and do not create value. But now even Socialist countries say that capitalists have three functions – starting new businesses, management and creativity.

"Most difficult is starting new businesses. The development of U.S. industry has relied on the constant formation of outstanding entrepreneurs. The Soviet Union killed capitalists and had no managers. So, its economy failed ... 'Historical materialism' denied the existence of social sciences. For a long time, the Soviet Union did not recognise this discipline. In China, it was only set up after the Cultural Revolution, more than 20 years later than the Soviet Union. History has no end. Communism was considered the highest stage [of history]. This was not logical.

"The Soviet Union collectivised agriculture and took over industry, leading to mass famine and purges ... During the 'Great Cultural Revolution' in China, Chinese starved to death. The number who died an unnatural death was 70 million. New research abroad puts the figure at 83 million.

In the Soviet Union, 60 million died an unnatural death, a higher proportion than in China. So, the Soviet Union collapsed on its own."

In 2011, he told National Public Radio of the U.S. that he hoped to live long enough to see the Chinese government admit that the bloody crackdown on Tiananmen Square's pro-democracy protests in 1989 had been a mistake. After the crackdown, Zhou resigned from the CPPCC in protest; he had served there since 1965, except during the Cultural Revolution.

Later, when he fell ill, he sought treatment at the 301 Hospital which cares for senior Party and government leaders, including scholars of his rank; it is the most prestigious hospital in Beijing. But the Party's Organisation Department declined to provide the needed pass; it needed the intervention of a senior doctor there who knew Zhou to obtain permission for him to enter.

Praise of Chinese Intellectuals

On January 13, 2013, the *New Beijing Daily* and *Finance Magazine* organised an event to celebrate Zhou's 108th birthday. The scholars present praised him as a "global citizen" and the "oldest pioneer of Chinese culture and thinking and the new enlightenment".

On January 16, 2016, friends held a party at his house to celebrate his 110th birthday. They praised him as one of China's outstanding intellectuals. One commented that Zhou had published three books at the ages of 100, 104 and 105.

Ma Guochuan, a scholar and author, said that the three books showed Zhou was a clear-headed personality of contemporary Chinese thinking, culture and history and an important enlightenment thinker. "His main belief is that China must advance toward modern civilisation. All countries are competing on the same path. For a period, the country may make a bend in the road but finally must go back to the common civilisation of mankind."

Li Weiguang, a professor at Tianjin University of Finance and Economics, said that Zhou advocated a global era. "We must see a country from the world, not see the world from a country. He does not promote 'a country' because in the world there is no 'country'. Learning is global and not divided by country. We hope that this wise and far-sighted old man can create a Guinness Book for his age."

Zhang Sengen, a historian and professor at the Chinese Academy of Social Sciences, said that Zhou was the oldest pioneer in the new enlightenment of Chinese thought and culture. "In the hearts of people, he is not only an outstanding scholar of linguistics but also a public intellectual who inspires respect. He is loved by middle-aged and old people and is loved and respected by young people. He is the pride of contemporary Chinese. He has a glorious and indisputable page in the history of Chinese and global culture." Peking University and 10 other universities established a new faculty, for the study of "Modern Chinese Characters".

Passing

At 03:00 on January 14, 2017, Zhou's condition deteriorated and he was taken to the emergency unit of the Beijing Union Hospital. At 04:00, his doctor said that he could not help him. He died that day; it was one day after his 111th birthday. His funeral was held on January 21 at a funeral home in an eastern suburb of Beijing, attended by family members and many friends.

His passing provoked an outpouring of grief. "You were my dear friend for decades," wrote Victor Mair, Professor of Chinese at the University of Pennsylvania. "I wish that you had gone on living forever. You will be sorely missed, but yours was a life well lived. As the 'Father of Pinyin', you have had an enormous impact on education and culture in China. After you passed the century mark, you spoke out courageously in favour of democracy and reform. Now, one day after your 111th birthday, you have departed, but you will always be in our hearts, brimming with light, as your name suggests." The word "guang", part of Zhou's name sound, means "shining".

Three years earlier, to mark his 108th birthday, Mair had written: "I wish to honour my ageless friend for his indomitable courage and brilliant acumen in confronting China's social and political challenges as resolutely and rationally as he tackled China's language issues over sixty years ago. His mind is still sharp as a tack, and every day he sits at his little desk to write articles and books on the tiny Sharp typewriter that he helped design. All of this is quite remarkable for a man who was born while the Manchus still ruled China. Even more astonishing for a man of his great age, Zhou Xiansheng (Mister) is arguably the most outspoken proponent of democracy and freedom of speech in China at a moment when such topics have become increasingly dangerous to broach."

Su Peicheng, a linguistics professor at Peking University, said: "Mr Zhou is a great scholar whose knowledge has spread from East to West. He was the founder of the theory of modern learning for the Chinese language."

On the day of his passing, Zhou was to have had an important meeting – with American Dr Vinton Gray Cerf, one of the founders of the Internet. In July 2016, Cerf had met in the United States a senior Chinese Internet official who told him of how Zhou and Pinyin had helped millions of

Chinese access the Internet. Cerf asked her to arrange a meeting with Zhou on his next visit to Beijing. The rendezvous was set for January 14, 2017. "I only wish to see him, shake his hand and take a photograph," said Cerf. "That would be perfect. It would be a historic meeting."

That morning, after completing his other appointments, Cerf was preparing to go to Zhou's apartment when he heard the sad news. He made a plaque, written in English and Chinese: "In memory of Zhou Youguang whose brilliant and persistent invention of Pinyin helped to bring the Internet and its applications within reach of the Chinese-speaking community. Long may he be remembered!" Cerf signed it "co-inventor of the Internet".

On October 26 that year, Zhou's ashes and those of his wife were taken to his home town of Changzhou. On the morning of November 2, the city government held a ceremony to bury the ashes in the Qifengshan Famous People's Garden in the city. The two had a tombstone 1.73 metres high. The event was attended by more than 60 people, including granddaughter Zhou Heqing and other family members, the vice-director of the Applied Language and Character Research Centre of the Ministry of Education, scholars and leading officials of the city. In 2015, Changzhou University established the Zhou Youguang Language and Culture College, with courses in English, Japanese, Spanish and Chinese.

Few scholars in history have made Zhou's contribution to the world. Pinyin has enabled more than one billion Chinese to learn their own language quickly and easily and turned their country into a literate nation. Similarly, it has helped tens of millions of foreigners to learn one of the world's most difficult languages.

Then, with the arrival of the electronic age, Pinyin has become an essential tool for Chinese and non-Chinese to access the Internet and other online platforms. I venture to say that, when Zhou arrived at the gates of Paradise, the angels opened the gates for him without the need of an interview or examination of his documents. Who among us has used his time on earth to such a good purpose?

Bibliography

1. Alan Simon.(2009, March 26).Father of Pinyin.*China Daily.*

2. Alpha History on the Great Leap Forward.Retrieved from https://alphahistory.com/chineserevolution/great-leap-forward/

3. A Meeting with Zhou Youguang.(2010).*World Affairs magazine,* issue #13.

4. Anonymous.(2017, January 15).Obituary of Zhou Youguang.*Ming Pao.*

5. Anonymous.(2017, January 24).Obituary of Zhou Youguang.*Hong Kong Economic Journal.*

6. Bank of Jiangsu: A local financial pioneer through a century. Retrieved from http://finance.china.com.cn/roll/20191223/5156288.shtml

7. Burial of ashes of Zhou Youguang in a garden in Changzhou. Retrieved from https://www.thepaper.cn/newsDetail_forward_1855266

8. Changzhou University website. Retrieved from http://eng.cczu.edu.cn/main.htm

9. Chen Guangzhong.(2012).*Follow and Read Zhou Youguang.*Chinese Creation Publishing Co., Ltd.

10. Chiung Wi-vun.(2013).Missionary Scripts on Vietnam and Taiwan. *Journal of Taiwan Vernacular,* 5(2), 94-123.

11. Encyclopaedia Britannica. Retrieved from https://www.britannica.com/topic/Encyclopaedia-Britannica-English-language-reference-work

12. Evolution of Internet in China,China Education And Research Network website.Retrieved From https://www.edu.cn/english/cernet/introduction/200603/t20060323_4285.shtml

13. Fan Yanpei, Zhang Sengen(2015).*Interviews with Zhou Youguang aged 110.*Cosmos Books Ltd.

14. Interview with Zeng Yi, Ph.D. candidate at Lingnan University, Hong Kong (2022, July 24).

15. John DeFrancis.(1950).*Nationalism and Language Reform in China.*Princeton University Press.

16. John DeFrancis.(2006).*The Prospects for Chinese Writing Reform.* Sino-Platonic Papers.

17. John King Fairbank and Merle Goldman.(1992).*China: a New History.*Belknap Press of Harvard University Press.

18. Jonathan D. Spence.(1990).*The Search for Modern China.*Century Hutchinson and W. W. Norton & Company

19. Kang-i Sun Chang.(2011, December).Zhang Chonghe and Overseas Kunqu. *Journal of the Studies of Ancient Texts,* Special issue for the Peking-Yale University Conference: Perspectives on Classical Chinese Texts and Culture (Beijing: Peking University Press), no. 11

20. Lin Peiyin.(2015). Language, Culture, and Identity: Romanization in Taiwan and Its Implications. *Taiwan Journal of East Asian Studies,* 12(24), 191-233.

21. Mao Xiaoyuan(niece of Zhou Youguang).(2015,October).I am a drop of water, Uncle is like a Great Ocean.

22. Mark O'Neill.(2010, July 20).Zhou Youguang calls it as he sees it – and he's not afraid to offend. *South China Morning Post.*

23. Mark O'Neill.(2010, May 30).How a linguist set out to rewrite Chinese history.*South China Morning Post.*

24. Richard W. Riby.(1980).*The May 30 Movement: Events and Themes.*Dawson Publishing.

25. Roxane Witke.(1977).*Comrade Chiang Ch'ing.*Wiedenfeld & Nicolson.

26. St. John's University.Retrieved from https://divinity-adhoc.library.yale.edu/UnitedBoard/St._ John's_University/StJohns-about.pdf

27. Tania Branigan.(2017, February 1).Obituary of Zhou Youguang.*The Guardian.*

28. The 1930s and War Economy.Retrieved from

 https://www.grips.ac.jp/teacher/oono/hp/lecture_J/lec09.htm

29. Thomas Bird.(2022, September 18).The Ascent of Mando.*South China Morning Post Magazine.*

30. Wang Xin.(2017, January 23).How Vinton Gray Cerf nearly met Zhou Youguang.*Bloomberg.*

31. Wen Huiwang.(2017, January 14).Granddaughter Zhou Heqing interview.Wen Hui Wang website.

32. Wu Hongfei.(2010, February 8).Zhou Youguang: At 105 from the World See China.*Southern People Weekly.*

33. Zhang Sengen.(2013, January 16).Evergreen Zhou Youguang.*South Reviews.*

34. Zhou Heqing(Zhou Youguang's Grand-daughter). (2017, January 14).Looking back on the life and personality of her grandfather.Wen Hui Wang website.

35. Zhou Youguang, 109 and Going Strong. Retrieved from https://languagelog.ldc.upenn.edu/ nll/?p=17097

36. Zhou Youguang, At 105, from the World See China(*Southern People Weekly*).Retrieved from https://www.infzm.com/contents/41515

37. Zhou Youguang,Chen Xiaoping.(2010, January 18).Zhou Youguang: Chinese Characters are a bottomless pit.*China News Weekly.*

38. Zhou Youguang.(2015).*My Hundred Years As I Told It.*The Chinese University of Hong Kong Press.

39. Zhou Youguang 1906-2017. Retrieved from https://languagelog.ldc.upenn.edu/nll/?p=30382

40. Zhou Youguang and His Era – The Century of History of One Man. Retrieved from https:// www.jiemian.com/article/1070662.html

41. Anonymous.(2010, July 20).Zhou Youguang calls it as he sees it – and he's not afraid to offend. *South China Morning Post.*

42. Anonymous.(2010, April 29).Zhou Youguang – Harmony is an Ideal, Modest Prosperity is Realistic.*Oriental Outlook.*

43. Zu Wei.(2016, January 11).Language Scholar Zhou Youguang 111 years old: "My age is old, my thinking is not old". *Beijing Youth Daily.*

44. "Modern Decline of Railroads".Retrieved from https://history.howstuffworks.com/american-history/decline-of-railroads.htm

THE MAN WHO MADE CHINA A LITERATE NATION –
ZHOU YOUGUANG, FATHER OF THE PINYIN WRITING SYSTEM

Author   Mark O'Neill
Editor   Donal Scully, Shirley Hau
Designer Vincent Yiu

Published by Joint Publishing (H.K.) Co., Ltd.
20/F., North Point Industrial Building, 499 King's Road,
North Point, Hong Kong

Printed by Elegance Printing & Book Binding Co., Ltd.
Block A, 4/F., 6 Wing Yip Street, Kwun Tong, Kowloon, Hong Kong

Distributed by SUP Publishing Logistics (HK) Ltd.
16/F., 220-248 Texaco Road, Tsuen Wan, N.T., Hong Kong

First Published in September 2023
ISBN 978-962-04-5281-9

三聯書店
http://jointpublishing.com

JPBooks.Plus
http://jpbooks.plus